CHEMISTRY RESEARCH AND APPLICATIONS

ESSENTIAL OILS

HISTORICAL SIGNIFICANCE, CHEMICAL COMPOSITION AND MEDICINAL USES AND BENEFITS

CHEMISTRY RESEARCH AND APPLICATIONS

Additional books in this series can be found on Nova's website
under the Series tab.

Additional e-books in this series can be found on Nova's website
under the e-book tab.

CHEMISTRY RESEARCH AND APPLICATIONS

ESSENTIAL OILS

HISTORICAL SIGNIFICANCE, CHEMICAL COMPOSITION AND MEDICINAL USES AND BENEFITS

MIRANDA PETERS
EDITOR

nova publishers
New York

Library of Congress Cataloging-in-Publication Data

Names: Peters, Miranda, 1975- editor.
Title: Essential oils : historical significance, chemical composition, and medicinal uses and benefits / editor, Miranda Peters.
Description: Hauppauge, New York : Nova Science Publishers, Inc., 2016. | Series: Chemistry research and applications | Includes bibliographical references and index.
Identifiers: LCCN 2015046169 (print) | LCCN 2016001085 (ebook) | ISBN 9781634843515 (hardcover) | ISBN 9781634843676 ()
Subjects: LCSH: Essences and essential oils. | Essences and essential oils--Therapeutic use.
Classification: LCC QD416 .E8765 2016 (print) | LCC QD416 (ebook) | DDC 615.3/21--dc23
LC record available at http://lccn.loc.gov/2015046169

Published by Nova Science Publishers, Inc. † New York

CONTENTS

PREFACE

The use of essential oils by civilizations constitutes a common practice since antiquity. In earlier times, China, India and the Middle East used herbs and oils in cooking, cosmetics, medicine and in religious rituals. These substances come from a secondary metabolism of plants and are associated with several functions necessary for their survival, such as the defense against microorganisms, predators and attraction of pollinators.

Essential oils are composed of a complex mixture of biologically active substances, lipophilic and volatile, and in most cases derivatives of terpene compounds and in a lower occurrence - phenylpropanoids. They have been long recognized for their medicinal uses: antibacterial, antifungal, antiviral, insecticidal and their antioxidant properties. The increased interest in alternative natural substances is driving the research community to find new uses and applications of these substances. This book provides research on the historical significance, chemical composition and medicinal uses and benefits of essential oils. Some of the topics discussed include extraction methods of essential oils; an overview of the most relevant essential oil changes regarding plant phenology and plant development; the antibacterial properties of some essential oils; essential oils applications in agriculture; and the use of essential oils as additives in broiler diets.

Chapter 1 - The use of essential oils has been a common practice in human societies since antiquity. The ancient civilizations of China, India, and the Middle East were already using herbs and oils in cooking, cosmetics, medicine, and religious rituals. These substances derive from plant secondary metabolism and are associated with several functions necessary for their survival, such as defense against microorganisms and predators and attraction of pollinators. Essential oils are composed of a complex mixture of biologically active substances, lipophilic and volatile, in most cases derivatives of terpene compounds and, to a lesser extent, phenylpropanoids. In recent years, interest in these substances has intensified due to the numerous properties associated with them, such as analgesic, expectorant, antimicrobial, antiseptic, insecticidal, fungicidal, and bactericidal qualities. These latter properties in particular have aroused the interest of many researchers in different parts of the world, seeking alternative products to replace synthetic insecticides and fungicides, due to the risks posed by the latter to health and ecosystems. The properties of essential oils make them good candidates for use in the production of medicines and natural insecticides.

Chapter 2 - The biosynthesis of the essential oils can be affected by a number of factors, mainly genetic and environmental ones. These potential variations in their chemical composition can occur and have been the subject of many researchers. Since many of the most commercially important species are perennial species, their harvest time and cultivation conditions can be upgraded through better knowledge of the variations in the phenology in order to improve yields and quality of their essential oil. Furthermore, examining the trend of qualities and yields over 5, 6 or even more years throughout the plant development provides a key tool for the strategic establishment of new promising species as alternative crops. Producers are interested in these alternative crops that may improve the environmental and economic sustainability of existing cropping systems. In most species, the chemical composition of the essential oils is mainly determined by individual variations and, therefore, is largely subjected to a genetic control. However, ontogeny changes throughout phenology can cause more or less pronounced variations in a few compounds of the whole essential oil chemical spectra. Notwithstanding these slight variations in chemical composition, the phenological stages have a greater influence on the yields obtained from distillation of dry material. The most recent published data leads to the suggestion that although flowering time is the most common harvest time for essential oil, producing plants may not always be in line with the highest yields. Herein, the authors provide an overview of the most relevant essential oil changes regarding plant phenology and plant development, summarize the latest reports related to the monitoring of essential oil production, and outline their implications for productivity and quality.

Chapter 3 - Essential oils (EOs) are natural volatile complex blends of biologically active molecules used nowadays for a panoply of applications. Presently, their importance have been highlighted due to the increasing demand by the food and pharmaceutical industries but also for their potential to provide therapeutic benefits in the prevention and management of diseases. In view of the multiple applications of the EOs, it is becoming important to explore different extraction techniques for higher yields of EOs as well as for the isolation of new biologically active compounds. This chapter attempts to highlight recent extraction techniques for obtaining EOs from medicinal aromatic plants. The conventional extraction techniques, their optimisation and improvements are discussed through the innovative principles, their benefits and disadvantages. Moreover, this chapter also endeavors to provide an up-to-date literature on EOs, the reported traditional usage of EOs from medicinal and aromatic plants of Mauritius, common extraction methods, bioactivities, and applications.

Chapter 4 - Bacteria are capable of causing severe infections, adapting to divergent situations and developing resistance to antibiotics. Products extracted from medicinal plants can be utilized as antibacterial agents and drug-resistance modifiers, thereby potentiating or mitigating the activity of synthetic antibiotics. The family Lamiaceae is characterized as having species with pharmacological potential, particularly with respect to combating bacterial infections. The aim of the present work was to evaluate the antibiotic and antibiotic-modifying activity of essential oil of *Ocimum gratissimum* Lamiaceae, using the microdilution technique. Standard and multiresistant bacterial strains of *Staphylococcus aureus, Escherichia coli and Pseudomonas aeruginosa* were utilized, and antibiotics of the aminoglycoside class were tested. With respect to evaluating the antibacterial effect of the *O. gratissimum* L, they were found to have a minimum inhibitory concentration (MIC) $\geq$ 1024 μg/mL, except *S. aureus* ATCC 25923 with MIC = 512 μg/mL. The drug-modifying potential of the essential oil of *O. gratissimum* L. was found to be antagonistic, reducing the effect of antibiotics, against the bacterial strains tested, except against *E. coli* when combined with amikacin and against *P. aeruginosa* combined with gentamicin, where there was no effect. The data obtained are promising, but further studies are needed to isolate the active compounds and to conduct pharmacological tests *in vivo*, making it possible to develop new therapeutic alternatives for the treatment of diseases caused by multiresistant microorganisms.

Chapter 5 - The essential oil from the leaves of *Plinia jaboticaba* was analyzed by CG-MS. The sesquiterpenoids constituted the dominant fraction in this oil. The major compounds found in this oil were sesquiterpenes spathulenol (17.29%), γ-murolene (13.7%) and valencene (10.35%). The monoterpene fraction correspond to 9.28% of this essential oil, being 1,8-cineole (6.01%) the compound present in the higher quantity in this class. This oil was assayed to determine its antibacterial activity against cell-wall bacteria *Staphylococcus aureus*, *Escherichia coli* and *Pseudomonas aeruginosa*, and against no-cell-wall bacteria *Mycoplasma arginini*, *Mycoplasma mycoides* subsp. capri (MMC) and *Ureaplasma urealyticum*. The oil of *P. jaboticaba* showed antibacterial activity against all tested bacteria, which the most accentuated for no-cell-wall bacteria giving values of MIC = 750 μg/mL.

Chapter 6 - Oxidative stress is defined as an imbalance between pro-oxidant and antioxidant species within the intracellular environment, where an overproduction of reactive oxygen species (ROS) may cause oxidative damage to cell structures, such as DNA, functional proteins or cell membrane lipids; the perpetuation of that situation finally leads to cell death. Extensive evidence demonstrate the key role of oxidative stress in the etiopathogenesis of neurodegenerative disorders, including Alzheimer´s and Parkinson´s diseases among others, which are characterized by the loss and dysfunction of nervous cells.

The use of natural exogenous antioxidants capable to counteract the oxidative damage via diverse mechanisms, such as scavenging of reactive species, potentiation of endogenous enzymatic defense or amelioration of mitochondrial dysfunction, attracts increasing research and emerges as a promising therapeutic tool. In this context, several natural products are currently being investigated as potential neuroprotector agents.

Essential oils are concentrated liquids with pleasant odors and composed of complex mixtures of volatile compounds that can be extracted from several plant organs by distillation processes. They are a good source of several bioactive compounds with pharmacological interest, and have been largely reported for antimicrobial, anti-inflammatory, anticancer and antioxidative properties. Some essential oils have been used in clinical medicine with different purposes, varying from hypnotic induction in sleep disturbances to the treatment of respiratory ailments.

The current chapter offers a general overview of the implications of oxidative stress in the neurodegeneration of Alzheimer´s and Parkinson´s diseases, with special focus on the antioxidant and neuroprotective properties demonstrated for essential oils. Results obtained in several neurodegeneration models are collected in this review and their therapeutic potential discussed.

Chapter 7 - Essential oils (EOs) have been long recognized for their medicinal uses: antibacterial, antifungal, antiviral, insecticidal and their antioxidant properties. The increased interest in alternative natural substances is driving the research community to find new uses and applications of these substances. Algeria, rich in indigenous herbal resources which grow on its varied topography and under changing climatic conditions permitting the growth of almost 3000 plant species (Cheriti et al. 2006). In the semi-arid regions, local traditional pharmacopoeia continues to be an important source of remedies for primary healthcare, so the region affords ample scope for studies concerning various aspects of folk medicine (Cheriti et al. 2006). The purpose of this chapter is to provide an overview on the variability of the chemical composition of the essential oil of some widely common used medicinal and endemic plants growing in Algeria: *Pistacia atlantica, Pituranthos scoparius* and *Rhanterium adpressum*. In the other hand, the growing interest in the substitution of synthetic food antioxidants by natural ones has fostered research on vegetable sources and the screening of raw materials for identifying new antioxidants. In this context, a survey of the antioxidant activities of theses oils is also reported. Furthermore, and due to the lack of a standard protocol of measuring antioxidant activity *via* DPPH assay, which is by the way the most employed *in vitro* technique, it is difficult to compare the results reported from different research groups; in this context, a new approach for effective comparison and expression is proposed.

Chapter 8 - Essential oils of plant species have shown activity against a variety of pathogenic and toxigenic fungi. The toxigenic fungi can cause a lot of damage to food commodities, including mycotoxin production. The presence of molds and mycotoxins in food commodities is a potential risk to health of humans and livestock. The aflatoxins, mutagenic and carcinogenic substances, are mycotoxins produced especially by *Aspergillus flavus* and *A. parasiticus*. This work reports the comparative effects of the three essential oils leaves on growth of A. flavus and their chemical constitutions. The essential oils of *Peperomia pellucida*, *Leunurus sibiricus* and *Porophyllum ruderale* were obtained by hydrodistillation in a Clevenger-type apparatus, and their effects were evaluated in *A. flavus* culture. *A. flavus* growth was evaluated by disk diffusion assay. Filter paper disks containing 0.0, 2.5, 5.0, 7.5, and 10.0 μL of essential oils were used, and commercial fungicide was the positive control. All volumes of the three essential oils reduced the fungal growth when compared with that of the control ($p < 0.05$). The essential oil of *P. ruderale* showed the greatest inhibition effect. *L. sibiricus* essential oil showed fungicidal effect greater than that of *P. pellucida* essential oil ($p < 0.05$). The chemical composition of the essential oils used was determined by GC-MS analysis. The use of essential oils with lower toxicity than those from synthetic products can be a good alternative to *A. flavus* control.

Chapter 9 - The overuse of chemicals in agriculture has caused several problems such as the emergence of resistant pests, environmental contamination, food residues as well as health hazards to the farmer. Insects, fungi and weeds cause losses in food production every year, by competing with crops and reducing its yield or by attacking products in post-harvest. In this way the search for new alternatives to control these agricultural pests is necessary. Essential oils are extracted from different structures of different plant species, and present compounds of the secondary metabolism of the plant that can have an effect on insects, fungi and weeds. Several bioactive compounds have already proven biological activity effect on several agricultural pests. Therefore the essential oils became a source for new products for the control of agricultural pests and help in the search for sustainable, environmentally friendly and socially fair agriculture.

Chapter 10 - The large production of poultry is in part due to antimicrobials that are used to improve broiler chicken performance, although in recent years this practice has been questioned due to suspected appearance of residues in the meat and resistant microorganisms. Thus, the essential oils emerge as promising substitutes to the usual growth promoters.

The aim of this study was to analyze the state of the art concerning the use of essential oils as additives in broiler diets and such data will be used to conduct further studies in the future.

The databases used were SciELO, Portal Capes, Science Direct and PubMed. 42 papers published between 2005 and 2014 were selected. 27 plant species were tested, and oregano was the most used. The essential oils act in different ways in the organism of the animals, going beyond the antimicrobial activity and showing effect on several productive parameters of the poultry, with results similar or better than those of antimicrobials.

Chapter 11 - The Ishpink tree, otherwise referred to as amazon cinnamon, *Ocotea quixos*, is a plant species that is part of the Lauraceae family. Ishpink is a tree, which is endemic to the Amazonian regions of Ecuador, Colombia and Peru. During the Spanish conquest, rumors of a legendary place full of gold and spices made the conquistadors venture into the impenetrable Amazonian forest. Due to the plant's described properties, it is extremely likely this was the species of legend.

Ishpink is an extremely aromatic plant species. Essential oil can be extracted from the bark, calyx and leaves. Various studies show chemical diversity within the oils, depending on which part it has been extracted from. Through these recent scientific studies, the plant's pharmacological properties are becoming appreciated.

Ecuadorian culinary practices utilize the plant's calyx as an ingredient in the preparation in one of the country's most iconic drinks "colada morada". This is a beverage saved for celebration during the day of the dead.

Recently, Ecuadorian community organizations such as Fundación Chankuap Recursos para el Futuro, have created small centres for the extraction and processing of Ishpink's essential oils. This approach is utilizing the plant's unique properties, translating them into a direct economic benefit for the local community.

The following research analyzes botanical, biological and chemical properties, historical and traditional uses, and potential economic and environmental benefits to the local communities through its sustainable exploitation.

In: Essential Oils
Editor: Miranda Peters
ISBN: 978-1-63484-351-5

Chapter 1

ESSENTIAL OILS: PROPERTIES, APPLICATIONS, EXTRACTION METHODS, AND PERSPECTIVES

Gustavo Haralampidou da Costa Vieira, Barbara Barbosa Dias and Denise Caren Ozório Leonel*
State University of Mato Grosso do Sul, Municipality of Cassilândia/MS, Brazil

ABSTRACT

The use of essential oils has been a common practice in human societies since antiquity. The ancient civilizations of China, India, and the Middle East were already using herbs and oils in cooking, cosmetics, medicine, and religious rituals. These substances derive from plant secondary metabolism and are associated with several functions necessary for their survival, such as defense against microorganisms and predators and attraction of pollinators. Essential oils are composed of a complex mixture of biologically active substances, lipophilic and volatile, in most cases derivatives of terpene compounds and, to a lesser extent, phenylpropanoids. In recent years, interest in these substances has intensified due to the numerous properties associated with them, such as analgesic, expectorant, antimicrobial, antiseptic, insecticidal, fungicidal, and bactericidal qualities. These latter properties in particular have aroused the interest of many researchers in different parts of the world, seeking alternative products to replace synthetic insecticides and fungicides, due to the risks posed by the latter to health and ecosystems. The properties of essential oils make them good candidates for use in the production of medicines and natural insecticides.

INTRODUCTION

Mankind's use of natural substances for mankind from the most primitive societies, in particular the Ancient Egyptian, Greco-Roman, and Chinese civilizations, where natural

* Corresponding author address: State University of Mato Grosso do Sul, Rodovia MS 306, km 6,4, Municipality of Cassilândia/MS, Brazil. E-mail: gcv@uems.br.

substances were used for a variety of purposes ranging from medicine to pest control (Machado and Fernandes Jr. 2011). In the literature there are records of the use of aromatic substances dating back more than four millennia, including in Chinese medicine and in medicinal and spiritual rituals in ancient Egypt (Stevensen 1998). Another classic example of the use of natural products for medicinal purposes, practiced from the earliest times to the present day, can be seen in India, where the aromatic plant *Syzygium aromaticum*, popularly known as clove, is used in Ayurvedic medicine to treat respiratory problems and eating disorders (Costa et al. 2011). Antiseptic and antibiotic properties are also attributed to this plant (Banerjee et al. 2006).

Among the natural products used for these purposes, essential oils are recognized for their pharmacological and therapeutic properties, as well as their low toxicity to mammals (Costa et al. 2011). These substances are also used in cooking to enhance the flavor of certain foods and beverages, for the inhibition of odors, and in the control of health problems (Franz 2010). The variety of different properties ascribed to essential oils has led to their use in a range of applications, including, among other things, as astringents, analgesics, antidepressants, antipyretics, antivirals, bactericides, bacteriostatics, deodorants, stimulants, fungicides, fungistatics, and insecticides (Worwood 1995).

Such properties derive from products formed through chemical reactions that occur continuously in the cells, directed by the action of enzymes, which make up the metabolism of plants. The synthesis of compounds common to living things and essential for the survival of plant species, such as sugars, amino acids, fatty acids, nucleotides and their derivatives, polymers, comprises the primary metabolism. On the other hand, compounds synthesized by other routes, and which appear to be no great utility in the survival of the species, are part of the secondary metabolism, which is characterized by the synthesis of compounds with diversity, structural complexity, production on a small scale, restricted distribution, and specificity. Essential oils are therefore referred to as secondary metabolites, being constituted by complex chemical elements which, as well as being associated with plant defense in the environment, are restricted substances in nature and limited to a smaller number of species (Morais 2009).

Essential oils, also known as volatile oils, are complex mixtures of lipophilic organic compounds, with low molecular weight, differing from fixed oils (lipid mixtures, usually from seeds, castor oil, cocoa butter, and flax seed oil) due to their highly volatile nature. Essential oils are relatively fluid, strong smelling, and poorly soluble in water. Typically colorless or slightly colored, they are stable in the presence of light, heat, and air, in addition to having a striking and pleasant odor (Morais 2009). These natural components are derived from aromatic and medicinal plants, abundant especially in the Myrtaceae, Rutaceae, and Lauraceae families, are arranged in the form of droplets between cells, and can act like hormones, regulators, and catalysts. Some reports in the literature describe their function as being to help the plant adapt to the environment, having stimulated production in stress situations. Some plants that live in very hot climates, such as in the desert, use essential oils to protect themselves from the sun. Another example is the tenuous cloud of essential oils formed around bushes of myrrh and frankincense in order to filter out the sun's rays and freshen the air around the plant (Lavabre 2009).

In addition to the properties mentioned earlier, recent studies have suggested that citrus oils, due to the presence in their composition of the terpene d-limonene, are anti-carcinogenic (Steffens 2010). According to this author, this compound acts by inducing the natural death of cancer cells or by inhibiting cell growth. Geraniol, an element found in some compositions of essential oils, also has inhibitory action on the proliferation of colon cancer cells, inducing the depolarization of the plasma membrane by interfering in ion channels (Machado and Fernando Jr. 2011). As for the biological properties attributed to essential oils, we can cite larvicidal (Rajkumar et al. 2010), antioxidant (Wannes et al. 2010), bactericidal (Silveira et al. 2012), analgesic and anti-inflammatory (Mendes et al. 2010), fungicidal (Barbosa et al. 2015), and antitumor activity (Silva 2008).

Empirical knowledge about essential oils has grown over time, and today it is estimated that 80 percent of the world's population frequently uses indigenous or traditional medicinal practices in addressing their primary health needs, especially those employing therapies that involve the use of herbal medicines (Bagetta et al. 2010). Sartoratto et al. (2004) point out that the use of medicinal plants still prevails in developing countries, especially in Asia, Africa, and Latin America, where there is a dependency on folk medicine, as an alternative solution to health problems. The use of these products in place of conventional medicine constitutes an alternative which obviates the problems caused both to ecosystems and to human health attributed to the overuse of synthetic products. Soylu et al. (2010) point out that, although the primary method of controlling phytopathogenic microorganisms continues to be through the use of synthetic products, their indiscriminate use can result in damage to the environment, such as contamination of natural resources and risks to human health, leading to the development of pathogen resistance, the elimination of beneficial species and biological ecosystem imbalance, in addition to the high costs associated with their application. Risks to human health and development of pathogen resistance also include to cases where the overuse of synthetics in the control of microorganisms causes diseases to humans.

India, China, Indonesia, and Brazil are considered the main producers of essential oils (Bizzo et al. 2009). The main centers of consumption are the United States, the European Union, France (the main importer), Japan, the United Kingdom, Germany, Switzerland, Ireland, China, Singapore, and Spain. It is believed that the annual market for essential oils is currently worth around $15 million, with an increase of approximately 11 percent per year.

As noted above, essential oils originate from the secondary metabolism of plants, constituted by complex chemical elements. According to Knaak and Fiuza (2010), the specific function of these oils in plants is still unknown: however, it is believed that during development, the higher plants synthesize terpenoids essential for growth, including gibberellin, pigments, and steroids. However, other classes of terpenes in plants have roles related to their volatility, acting on attraction of insects and other fertilizing agents, in protection from predators, pathogens, water loss, and increased temperature, and carrying out ecological functions as germination inhibitors. In aromatic plants, essential oils are easily found in the flowers, leaves, bark, rhizomes, and fruits (Bizzo et al. 2009).

They consists primarily of organic compounds of the phenylpropanoid family, mono-and sesquiterpenes, and other volatile components, with a predominance of terpenes (Franz 2010). The various biological properties ascribed to terpenes and phenylpropanoids include, *inter alia*, antimicrobial, fungicidal, and anti-interference properties (Oliveira et al. 2011). Monoterpenes belong to the most abundant and potent group of naturally-occurring substances (Knaak and Fiuza 2010). According to Knaak and Fiuza, essential oils are made up of more than 60 individual components, including terpenics, simple alcohols and terpenic hydrocarbons, aldehydes, ketones, esters, and phenols, among others. The most common are the monoterpenes linalool, geraniol, thujone, camphor, limonene, and others. Among the most common sesquiterpenes are farnesol, nerolidol, and bisabolene (Simões et al. 2003). It is known that the chemical components of the volatile oils are divided into two classes. The first, based on biosynthesis, comprises derivatives of terpenoids, formed by means of mevalonic acid-acetate. The second, comprising phenylpropanoid derivatives, are aromatic compounds formed by the shikimic acid pathway (Strapazzon 2004).

Numerous studies have elucidated the components present in essential oils thought to be responsible for their antimicrobial properties. The essential oil of rosemary is constituted from the elements alpha-pinene and camphene (Angioni et al. 2004), which gives it antimicrobial properties against bacterial and yeast strains (Schelz and Hohmann 2006). Menthol, menthofuran, terpinene, menthone, limonene, cineole, phellandrene and alpha-pinene are the main substances found in the essential oil of peppermint (Zago et al. 2009). The essential oil of lemongrass (*Cymbopogum citratus*) is composed of myrcene, geranial, and neral, in addition to other substances used in folk medicine for treatment of colds, dysentery, and headaches, and as tranquilizer and antispasmodic (Pereira et al. 2004), as well as applications utilizing its antibacterial functions (Nguefack et al. 2004). Ginger (*Zingiber officinale*) is rich in terpenes, and, as well as being very much appreciated as a spice, is also used, especially by the Chinese, in medicines for the treatment of dysentery, malaria, rheumatism, and influenza (Sabulal et al. 2006). The thymol and carvacrol present in the essential oil of oregano cause distortion of the physical structure of the cell resulting in destabilization of the membrane, altering its permeability, denaturing enzymes, and modifying the proton motor force through variations in pH and electrical potential (Burth 2004). The essential oil of cinnamon (*Cinnamomum zeylanicum*), found in the leaves of the plant, contains 23 constituents, with a predominance of eugenol (60%). The branches contain 36 substances, with a predominance of monoterpenes α and β-pinene, α-felandreno, p-cymene, limonene, linalool, α-sesquiterpene, copaene, β- caryophyllene, caryophyllene and alilbenzenos oxide, cinnamaldehyde, and cinnamyl acetate (Lima et al. 2005). This essential oil is used as a flavoring agent and natural preservative for food, in addition to having proven antifungal (Lima et al. 2006) and bactericidal (Matan et al. 2006) action. Cinnamon's main constituent, eugenol, is also the principal constituent of cloves (*Caryophyllus aromaticus*). Found in 80 to 90 percent of their composition (Silva 2009), it is recognized for its antibacterial and antifungal properties, and is also the most active when tested on lineages of *Escherichia coli* (Burt and Reinders 2003). The eugenol induces inhibition of amylase and protease by cells, in addition to the decay and lysis thereof.

The abovementioned properties ascribed to these substances, coupled with their low residual power and non-toxicity to mammals, make essential oils a product of great scientific interest, currently being widely exploited by the pharmaceutical and cosmetics industries, as well as in agricultural products, food production, and elsewhere.

With regard to food in particular, Runyoro et al. (2010) tell us that, despite advances in sanitation techniques, the contamination of these products by undesirable organisms during processing, storage, and distribution has been observed both in developing countries and 'first world' nations. With microorganisms' increased resistance to preservatives, the development of alternative products capable of inhibiting the development of pathogens is becoming increasingly necessary, and essential oils hold significant potential in this respect (Militello et al. 2011). Militello et al. observed that recent years have seen a growing interest among consumers for food products of organic or natural origin, free from any chemical residue. As consumers are already accustomed to using spices to enhance the flavor of foods, the use of essential oils derived from the same source for the purpose of inhibiting the development of pathogenic microorganisms would not be controversial. Recent years have seen an empirical increase in the incidence of food-borne diseases, and this issue is a matter of public health relevance (Oussalah et al. 2007).

One of the main microorganisms which has caused concern among health officials across the globe, and which constitutes a major barrier to international trade in food, are bacteria of the genus *Salmonella* (Bona et al. 2012). Its widespread distribution among animals and its capacity to remain in the environment for long periods of time mean that this zoonosis has important ramifications with regard to public health (Butaye et al. 2003). With regard to salmonellosis, Santurio et al. (2007) determined the effect of the essential oils of oregano (*Origanum vulgare*), thyme (*Thymus vulgaris*) and cinnamon (*Cinnamomum zeylanicum*) on 60 samples of *Salmonella enterica*, noting that the essential oil of oregano presented strong antibacterial activity, with a minimum inhibitory concentration (MIC) average 529 $\mu g.mL^{-1}$, followed by thyme, which presented a moderate activity (average 961μg CIM.mL^{-1}), while cinnamon was the substance which presented the least inhibitory effect (average CIM 1335$\mu g.mL^{-1}$). In another study, Santurio et al. (2011) determined the bactericidal effect of the essential oils of *Origanum vulgare* (oregano), *Thymus vulgaris* (thyme), *Cinnamomum zeylanicum* (cinnamon), *Lippia graveolens* (Mexican oregano), *Zingiber officinale* (ginger), *Salvia officinalis* (Sage), *Rosmarinus officinalis* (rosemary) and *Ocimum basilicum* (basil) in doses ranging from 400 to 6400 $\mu g.mL^{-1}$ against 79 samples of *Escherichia coli* isolated from the feces of birds (43) and cattle (36). They reported that the essential oils of oregano, Mexican oregano, thyme, and cinnamon present bactericidal activity on this microorganism, with oregano proving particularly effective.

Silveira et al. (2012) determined the bactericidal effect of the essential oils of basil (*Ocimum basilicum*), lemongrass (*Cymbopogon flexuosus* and *Cymbopogon winterianus*), cinnamon (*Cinnamomum zeylanicum*), fennel (*Foeniculum vulgare*), laurel (*Laurus nobilis*), mint (*Mentha arvensis*), pennyroyal (*Mentha pulegium*), orange (*Citrus sinensis*), oregano (*Origanum vulgare*), and rosemary (*Rosmarinus officinalis*) against the bacteria *Staphylococcus aureus, Lactobacillus plantarum, Enterococcus faecalis, Listeria monocytogenes, Bacillus cereus, Bacillus subtilis, Escherichia coli, Salmonella typhimurium, Proteus vulgaris, Enterobacter aerogenes, Pseudomonas aeruginosa*, and *Yersinia enterocolitica*. The ranked in descending order according to their bactericidal activity were lemongrass (*Cymbopogon flexuosus*), Basil (*Ocimum basilicum*), oregano (*Origanum vulgare*), cinnamon (*Cinnamomum zeylanicum*), and Laurel (*Laurus nobilis*). According to these authors, the results showed that the evaluated essential oils present high potential the natural preservatives.

In studies conducted with the bacteria *Staphylococcus aureus* and *Escherichia coli* isolated from human clinical cases, Silva et al. (2009) determined the *in vitro* bactericidal action of the essential oils of rosemary (*Rosmarinus officinalis*), clove *(Caryophyllus aromaticus)*, ginger *(Zingiber officinalis)*, lemongrass (*Cymbopogon citratus*), peppermint (*Mentha piperita*), and cinnamon (*Cinnamomum zeylanicum Blume*), reporting that cinnamon oil was the most efficient, showing minimal inhibitory concentration values equal to 0.047 and 0.09 for *S. aureus* and *E. coli*, respectively. The other studied oils were efficient, albeit with more discrete effects on these organisms.

Carson et al. (2006) argue that the antimicrobial action attributed to essential oils is linked to their ability to cause lysis and loss of integrity of the cell membrane due to the removal of ions. Other authors claim that essential oils can cause bacterial wall degradation, affecting the plasma membrane and membrane proteins, interfering with the flow of electrons and the coagulation of the cytoplasm (Ultee et al. 2002). Dorman and Deans (2000) presented a more complete explanation of the antimicrobial effect of essential oils, stating that these substances affect the structure of the bacterial cell wall, causing protein denaturation and coagulation, alter the permeability of the plasma membrane to hydrogen ions and potassium, harm the vital processes of the cell such as electron transport, translocation of protein phosphorylation and other dependent enzyme reactions, and thus affect the chemiosmotic control of the cell leading to mortality of bacteria. Sartoratto et al. (2004) attribute all these destructive processes observed in bacteria to monoterpenes present in vegetables.

In addition to the bactericidal action, essential oils can cause a synergistic effect when associated with medicines. The association between the derivative compounds present in medicinal plants and antimicrobial drugs is known to be able to inhibit or enhance the therapeutic effect of antimicrobial drugs, or even interfere with the expected response (Nascimento et al. 2000). Canton and Onofre (2010) studied the effect of essential oil of *Baccharis dracunculifolia* on 29 antibiotics used in clinical medicine, in most cases observing either synergic or antagonistic interferences of the antibiotic action. These authors suggest that the use of derived products or plants can interfere with the effectiveness of the antibiotics use in clinical treatment. Zago et al. (2009) studied the interactions between the aromatic plants cinnamon (*Cinnamomum zeylanicum*), lemongrass (*Cymbopogon citratus*), mint (*Menth piperita*), ginger (*Zingiber officinale*), clove (*Caryophyllus aromaticus*), and rosemary (*Rosmarinus officinalis)* and eight antimicrobial drugs against twelve *Staphylococcus aureus* and *Escherichia coli* strains isolated from human specimens. In this study, *S. aureus* was negatively influenced by oils and drugs associations, and the lemongrass oils presented synergism with all drugs. In *E. coli* assays, synergism occurred only with rosemary (three drugs) and lemongrass (two drugs).

In the literature, there are also reports of the fungicidal effect of essential oils. According to Espinel-Ingroff et al. (2005), the microorganisms *Candida* spp. and *Aspergillus* spp. are associated with approximately 80 to 90% of fungal infections. This pathogen produces several infections with invasive processes. *Aspergillus fumigatus* is one of the major causes of contamination to intrahospital environments (Haiduven 2008). For both species, *Candida* spp. (Cantón et al. 2008) and *A. fumigatus* (Snelders et al. 2008) there are cases of drug resistance. Correa-Royero et al. (2010) studied the fungicidal effect of 32 essential oils and extracts, 29 against *Candida krusei* and *Aspergillus fumigatus*, noting that the essential oil of *Lippia citriodora* does not affect this microorganism. Moreover, the essential oil of *Chenopodium ambrosioides* and the extract of *Myrcia cucullata* inhibited *C. krusei* in vitro at similar concentrations as fluconazole. Another work that evaluated the effect of different essential oils on the fungus genus *Candida* was conducted by Lima et al. (2006). These authors established the minimum inhibitory concentration (MIC) of aromatic plants *Cinnamomum zeylanicum*, *Citrus limon*, *Eucalyptus citriodora*, *Eugenia uniflora*, *Peumus boldus*, and *Rosmarinus officinialis* essential oils against *Candida albicans*, *Candida guilliermondii*, *Candida krusei*, *Candida parapsilosis*, *Candida stellatoidea*, and *Candida tropicalis* strains, in doses of 0.25, 0.5, 0, 1, 2, and 4. *C. zeylanicum* and *P. peumus* essential oils were the most efficient, inhibiting the growth of 58 of the be yeasts strains, and presented MIC of 4%.

Experiments performed by Barbosa et al. (2015) likewise reported the fungicidal effect of essential oils. These authors verified the fungitoxic effect of the essential oils of palmarosa (*Cymbopogon martinii*), tea tree (*Melaleuca alternifolia*), clove (*Eugenia caryophyllata*), and eucalyptus (*Corymbia citriodora*) in doses of 25, 50, 75, 100 and 125 μL.L^{-1} on the fungus *Colletotrichum musae*, which causes anthracnose in plantain. The authors noted that essential oils of tea tree, clove, and palmarosa in the dose of 50 μL.L^{-1} were effective in the control of this pathogen, with similar results to those obtained with the fungicide mancozeb at a dose of 90 g.L^{-1}. In a similar work, Carnelossi et al. (2009) evaluated the in vitro and in vivo effect of essential oils of *Cymbopogon citratus, Eucalyptus citriodora, Mentha arvensis*, and *Artemisia dracunculus*, in doses ranging from 1, 5, 10, 15, 25, 50 μL, on *Colletotrichum gloeosporioides*, a fungus that causes anthracnose in papaya. The authors noted that, in the largest dose tested, all of the oils tested were efficient in the control of the pathogen, and the *C. citratus* oil completely inhibited the mycelial growth of the fungus from the dose of 10 μL. Costa et al. (2011) determined the fungicidal effect of the essential oil of *Syzygium aromaticum* on hyphae of certain phytopathogenic fungi, reporting that the essential oil of cloves to 0.15 presents fungicidal activity on the fungi *Rhizoctonia solani*, *Fusarium solani*, *Fusarium oxysporum*, but that it does not interfere with *Macrophomina phaseolina* fungus mycelial growth.

Santos et al. (2013) investigated the in vitro fungotoxicity of the essential oils of lemongrass (*Cymbopogon citratus*), citronella (*Cymbopogon nardus*), lemon balm (*Lippia alba*), and peppermint (*Mentha piperita*), with five concentrations of essential oils (250 ppm, 500 ppm, 750 ppm, 1000 ppm and 1250 ppm) on the fungus *Helminthosporium* sp., observing that essential oils of lemongrass and citronella reduced mycelial growth of *Helminthosporium* sp., while all of the oils in their respective doses exhibited a preventive and curative effect.

In the case of fungi, the fungicidal activity of these substances is linked to their hydrophobicity, since it is the interaction between the substance and lipids of the cell wall, altering the permeability, that promotes alterations in these structures (Costa et al. 2011). According to Pascholati and Leite (1995), the accumulation of phytoalexins, the result of secondary metabolites, antimicrobials of low molecular weight produced by plants, and pathogenesis-related protein, cause degradation of the cell wall of fungi. One of the studies showing the effect of essential oils on fungi was that of Soylu et al. (2010), who observed morphological changes in the hyphae of *Botrytis cinerea* when exposed to oregano essential oil. This cytotoxic effect represents an important alternative method of controlling phytopathogens in agriculture (Bakkali 2008).

These works suggest the promising potential of essential oils in the control of microorganisms, both in agriculture and medicine, as an alternative to conventional treatments, which are often responsible for unwanted effects such as the development of pathogen resistance or side-effects. In agriculture, other factors such as the high cost of implementation and the contamination of the environment should be considered.

Insecticidal effects have been recognized in essential oils from citrus fruits. In insects, these substances can act in a number of different ways, causing toxicity (Hiremath et al. 1997), slowing development, inhibiting feeding (Wheeler and Isman 2001), causing the deterrence of oviposition (Zhao et al. 1998), and reducing fecundity and fertility (Muthukrishnan and Pushpalatha 2001).

One of the most significant studies on the effect of essential oils on insect mortality was conducted with *Triatoma infestans* (Hemiptera: Reduviidae), the transmitter of Chagas disease. According to data from the World Health Organization (WHO 2010), 10 million people across the world are infected by Protozoan *Trypanossoma cruzi*, the majority of them in Latin America, where the disease is endemic and constitutes one of the main public health problems. In recent decades, this disease has been diagnosed in the United States, Canada, various European countries, and certain countries of the Western Pacific, due to processes of human migration between Latin America and other countries. More than 25 million people worldwide are at risk of the disease (Gomes and Favero 2011). These authors evaluated the insecticidal action of essential oils of *Anacardium humile*, *Cymbopogon nardus*, *Ocimum basilicum*, and *Eucalyptus urograndis* in two application methods, topic action and steam pressure (fumigation), on *Triatoma infestans* nymphs, and observed that the essential oils of *O. gratissimum* and *E. urograndis* showed insecticidal activity against *T. infestans*, while *A. humile* and *C. nardus* showed low average mortality, 5 to 15% respectively. For the fumigation, only the *E. urograndis* oil caused 100% mortality on *T. infestans*.

Pereira et al. (2008) determined the insecticidal effect of essential oils of *Cymbopogon martini*, *Piper aduncum*, *Piper hispidinervum*, *Melaleuca* sp., and *Lippia gracilis*, and fixed oils of *Helianthus annuus*, *Sesamum indicum*, *Gossypium hirsutum*, *Glycine max* and *Caryocar brasiliense*, in doses of 0.5, 1.0, 1.5, 2.0 and 2.5 L.tons^{-1}, on the weevil, *Callosobruchus maculatus*, in caupi beans always green. The authors reported that the essential oils of *C. martini*, *P. aduncum* and *L. gracilis* caused 100% mortality in all concentrations studied, *P. hispedinervum* from 1.5 L. tons^{-1}, and *Melaleuca* sp. in concentrations of 2.0 and 2.5 L. tons^{-1}. These substances reduced the number of viable eggs and led to around a 100% reduction in the number of insects that emerged; the one exception to this was *Melaleuca* sp., the fixed oils, which presented low mortality in all tested concentrations, although reducing to almost 100 the number of viable eggs and insects that emerged. It is believed that the essential oil added to seeds has an insecticidal function, mainly due to the presence of triglycerides, in addition to causing the blockage of oxygen. Another advantage observed with the use of these substances in the treatment of stored grain is that they do not render the seeds and leave no residues (Knaak and Fiuza 2010).

Lima et al. (2009) investigated the influence of *Psidium guajava* oil diluted in ethanol and water 1:1 at a concentration of 0.01 and 0.001 on the behavior of caterpillars of the fall armyworm *Spodoptera frugiperda*, reporting that this essential oil at a concentration of 0.01 has a repellent effect on these insects. These authors characterized this oil chemically, reporting the presence of α-terpineol (0.9), 1.8-cineole (7.0), beta-caryophyllene (7.2) and caryophyllene oxide (13.8).

Soares et al. (2011) evaluated the insecticidal effect of essential oils extracted from *Piper hispidinervum*, *Cymbopogon citratus*, *Citrus limon*, *Syzygium aromaticum*, and *Rosmarinus officinalis* at concentrations of 1.0, 5.0 and 10.0 (vv) on the looper *Thyrinteina arnobia* eucalyptus brown, noting that *Citrus limon* alone was not efficient, while all other oils studied proved to be promising insecticides against *T. arnobia*.

In another work, Soares et al. (2012) quantify the constituents of essential oils from dry fruits of *Illicium verum* at concentrations of 0.3 to 2.0 (vv) and *Piper hispidinervum* at concentrations of 1.0 to 2.5 (vv), and its effects on the population of the aphid *Macrosiphum euphorbiae*, a parasite of over 200 species of plants from more than 20 families. According to the authors, *I. verum* oil has a high concentration of the compound (E)-anethole, and shows higher insecticide action than the oil of *P. hispidinervum*.

Lastly, Vieira et al. (2012) investigated the acaricidal effect of different essential oils against the *Varroa destructor* acarus, considered the major bee pest worldwide. This study tested the oils of anise, rosemary, cinnamon, eucalyptus, clove, and mint in concentrations of 10 $\mu L.L^{-1}$, 50 $\mu L.L^{-1}$, and 200 $\mu L.L^{-1}$. The anise, cinnamon, eucalyptus, and clove oils caused a significant mortality rate for the acarus. For the 200 $\mu L.L^{-1}$ concentration, the mortality rates reached 92.5 for anise and eucalyptus, 52.5 for cinnamon, and 87.5 for clove, suggesting that these substances can replace conventional acaricides. The study reported that the essential oils under investigation had no effect on the bees.

In the case of insects and mites, Prates and Santos (2002) conclude that plants with insecticidal properties are rich in secondary compounds, especially monoterpenes and the like, which, according to their lipophilic properties, influence the basic biochemical processes so as to induce physiological and behavioral effects in these invertebrates. According to Isman (2006), essential oils can interact with the insect integument, in addition to acting in digestive enzymes and neurological disorders. Viegas Jr. (2003) says that the terpenic compounds present in essential oils may act by inhibiting the action of the enzyme acetylcholinesterase, causing the accumulation of the neurotransmitter in the synaptic cleft, thereby preventing the correct transmission of the action potential, resulting in the death of the insect by respiratory failure (Chambers and Carr 1995). The our, substance found in nim, acts on the chemoreceptors, stimulating the cells' 'specific deterrents' or blocking the phagostimulants, thus resulting in inhibition of the insects' feeding (Mordue and Nisbet 2000). Knnak and Fiuza (2010) point out that the search for new natural products with insecticidal activity is a promising path, given the importance of phytotoxins to the development of alternative bioinsecticides.

As mentioned earlier, essential oils are also recommended in therapeutic treatments. In aromatherapy, the use of these substances induces improvements in mood, reduces stress induced by anxiety, depression, and chronic pain, the inhalation of volatile components having therapeutic functions in respect of both psychological and physiological effects (Bagetta et al. 2010). According to Jäger et al. (1992), Lavender oil (*Lavandula* spp.), one of the principal essential oils used in aromatherapy, is recommended in inhalations and baths in cases of nervous tension, rheumatism, and skin diseases such as eczema and dermatitis. Lavender oil is also used to promote sleep and relieve stress.

Another oil used in the treatment of skin complaints, respiratory problems, and hormonal issues through the practice of aromatherapy is the geranium (*Pelargonium* spp.) (Jalali-Heravi et al. 2006). Corazza (2004) reports that the geranium's use in aromatherapy is justified on the basis of their psychological and emotional actions as regards relaxation and relief of nervous tension, anxiety, and depression. Among other medicinal properties, this oil has been recognized for its analgesic action and capacity to regulate female hyposecretion of estrogen, as well as its diuretic, hemostatic, and cicatrizant properties, and is prescribed for alleviation of the symptoms of menopause and acne.

Experiments performed by Bagetta et al. (2010) showed that the components present in essential oils can enter the bloodstream and cross the blood-brain barrier, affecting the central nervous system through various channels, and may be administered orally, in the form of inhalation, through the skin (topical), or via subcutaneous or intraperitoneal injection (Orafidiya et al. 2004). Tanida et al. (2005) studied the effect of essential oils on the central nervous system in rats, noting that the components of these substances increase the activity of the sympathetic nerves and suppress the activity of the vagus nerve, which corresponds to a parasympathetic nerve, thereby increasing levels of glycerol in plasma, raising the blood temperature, and decreasing appetite. The penetration capability of the essential oils in the body through the skin has been proven in a study by Jäger et al. (1992). These authors observed that peanut vegetable oil plus lavender essential oil to presented percutaneous absorption in humans, after five minutes of massage. This study also detected in the blood traces of linalool and linalyl acetate, major components of lavender oil; after 20 minutes, maximum concentrations were obtained of these constituents in the plasma, i.e., 100 $ng.mL^{-1}$ of linalyl acetate and 121 $ng.mL^{-1}$ of linalool. Thus, they concluded that the relaxing effect of lavender oil occurs by two mechanisms, either through penetration through the skin or by inhalation.

Another important aspect regarding the use of essential oils relates to the method by which they are obtained, as the chemical composition of the oil is related to this, alongside other factors such as the weather, season, geographical conditions, and harvest period (Maciel et al. 2002). Essential oils can be extracted through various techniques, such as steam distillation, hydrodistillation, cold pressing or expression, enfleurage, extraction by organic solvents, and supercritical CO_2 extraction (Okoh et al. 2010).

Steam distillation is the most widely used extraction process. This method entails placing the plant material in the distiller, which, through the passage of steam through the plant material, extracts the volatile aromatic compounds from the plant; the compounds pass through the condensation system and are collected in a decanting container, where the water separates itself naturally from the oil thus formed, and the oil is removed from the container through a tap. The oil, thus obtained, is placed in a decanting funnel to ensure a thorough separation of water, before being bottled in amber glass and stored in a place sheltered from high temperatures and luminosities.

Hydrodistillation is the most widely used method on a laboratory scale. In this method, the vegetable raw material remains in contact with boiling water; the steam forces the opening of cell walls and oil evaporation occurs between plant cells. The steam, which consists of the mixture of oil and water, then passes through a condenser where it cools, and, as the volatile components and water are immiscible, the formation of two liquid phases that can be separated takes place (Silva 2011).

The oldest method of obtaining of essential oils is probably extraction through the use of fats. The extraction technique known as enfleurage consists of a layer of fresh flower petals, usually on a surface (a glass plate or cloth) in contact with vegetable or animal fat, which will act by absorbing the oil in question. Dead flowers are replaced with new ones every 24 hours, until the desired concentration is obtained. This technique is used for the extraction of more unstable essential oils, which may lose their aromatic components if extracted by other methods, as well as being widely applied to the extraction of essential oils from flowers, mainly for the perfume market (Lavabre 2009).

Pressing or cold expression is the most widely-used method for extraction of essential oils from citrus fruits. In this process, the fruits are placed in one piece directly in a hydraulic press, which collects the juice and oil present in the bark (Pinheiro 2003). After pressing, the oil is separated from the emulsion formed with water through decantation, centrifugation or fractional distillation (Simões et al. 2003).

Some types of oils are very unstable, and intolerant of temperature elevations. In these cases, organic solvents can be used for their extraction, such as hexane, benzene, methanol, ethanol, propanol, acetone, pentane, and a number of chlorinated solvents (Fillipis 2001). In the process of solvent extraction, the plants are immersed in a suitable organic solvent. After an interval of time sufficient to allow the transfer of soluble constituents present in the plant, the separation of the solid and liquid phases occurs. The oil is obtained through evaporation of the solvent present in the liquid phase (Steffani 2003).

Supercritical carbon dioxide (CO_2) extraction enables high quality oils to be obtained, since no trace of solvent remains in the final product, making it purer than those obtained by other methods. A supercritical fluid is one in which the gas is at a temperature at which it cannot be liquefied by isothermal compression. The temperature at which this occurs is known as the critical temperature. When a gas is in a condition where both the pressure and temperature are at levels higher than critical values the gas is said to be in its supercritical state. Under these conditions, the compressed gas has low viscosity (comparable to that of gases) and high density (high ability of dissolution, akin to that of liquids), allowing the use of the same in extraction processes of solutes from solid matrices. For such extraction, CO_2 is taken to the supercritical fluid state using conditions where the pressure is up to 200 atmospheres and a temperature of 33°C. Once the extraction has been performed, with the correct balance between the pressure of the substance and the pressure of the environment, recovery causes the CO_2 to return to its gaseous state, resulting in its total elimination (Steffani 2003).

In general, in addition to their numerous other properties presented in this work, it has been observed that essential oils have low toxicity to mammals and relatively low cost. These products, from vegetable raw materials, offer a wide variety of molecules with biological activity which can be precursors for the synthesis of new products (Duke et al. 2000). Many of these substances are already employed in modern pharmacology, since many of the synthetic drugs currently on the market are derived, directly or indirectly, from medicinal plants (Silver and Bostian 1993). There are also records of the use of commercial products based on essential oils for pest control in Europe and the U.S. (Knaak and Fiuza 2010). Given the speed with which microorganisms and pests have developed resistance to conventional chemicals employed to control them, research into new chemical groups with bactericidal, fungicidal, and insecticidal properties is becoming increasingly necessary. It is believed, especially in light of the adverse effects associated with the overuse of synthetic substances, both in conventional medicine and agricultural production, that the use of products from medicinal plants provides a promising path to the success of modern medicine, as well as being indispensable in pest management programs and in the pursuit of sustainable development.

REFERENCES

Angioni, Alberto, Barra, Andreia, Cereti, Elisabetta, Barile, Daniela, Coisson, Jean D., Arlorio, Marco, Dessi, Sandro, Coroneo, Valentina and Cabras, Paolo. 2004. Chemical composition, plant genetic differences, antimicrobial and antifungal activity investigation of the essential oil of *Rosmarinus officinalis* L. *Journal Agricultural Food Chemistry* 52(11): 3530-535. Doi: 10.1021/jf049913t.

Bagetta, Giacinto, Morrone, Luigi A., Rombolà, Laura, Amantea, Diana, Russo, Rossella, Berliocchi, Laura, Sakurada, Shinobu, Rotiroti, Domenicantonio and Corasaniti, Maria T.

2010. Neuropharmacology of the essential oil of bergamot. *Fitoterapia* 81(6): 453-61. Sep. 2010. doi: 10.1016/j.fitote.2010.01.013.

Bakkali, Fátima, Averbeck, Simone, Averbeck, Dietrich and Idaomar, Mohammed. 2008. Biological effects of essential oils – A Review. *Food and Chemical Toxicology* 46(2): 446-75. Doi: 10.1016/J.FCT.2007.09. 106.

Banerjee, Sarmistha, Panda, Chinmay K. and Das, Sukta. 2006. Clove (*Syzygium aromaticum* L.), a potencial chemopreventive agent for lung cancer. *Carcinogenesis* 27(8): 1645-54. Doi: 10.1093/CARCIN/BGI372.

Barbosa, Marcelo S., Vieira, Gustavo H. C. and Teixeira, Anielli V. 2015. Biological activity *in vitro* propolis and essentials oils on the *Colletotrichum musae* isolated in *Musa* spp. *Brazilian Journal of Medicinal Plants* 17: 254-261. doi.org/10.1590/1983-084X/13_063.

Bizzo, Humberto R., Hovell, Ana Maria C. and Rezende, Claudia M. 2009. Essential oils in Brazil: general aspects, developments and prospects. *New Chemical* 32(3): 588-94. doi.org/10.1590/S0100-40422009000300005.

Bona, Tânia D. M. M., Pickler, Larissa, Miglino, Leonardo B., Kuritza, Leandro N., Vasconcelos, S. P. and Santin, Elizabeth. 2012. Essential oil of oregano, rosemary, cinnamon and pepper extract in control the *Salmonella*, *Eimeria* and *Clostridium* in chicken. *Search Brazilian Veterinary* 32(5): 411-18.doi.org/10.1590/S0100-736X2012000500009.

Burt, Sara A. and Reinders, Robert D. 2003. Antibacterial activity of selected plant essential oils against Escherichia coli O157:H7. *Letters Applied Microbiology* 36: 162-67. Doi: 10.1046/j.1472-765X.2003.01285.x.

Burt, Sara. 2004. Essential oils: Their antibacterial properties and potential applications in foods – a review. *International Journal of Food Microbiology* 94: 223-253. doi:10.1016/j.ijfoodmicro.2004.03.022.

Butaye Patrick, Devriese Luc A. and Haesebrouck Freddy. 2003. Antimicrobial growth promoters used in animal feed: effects of less well know antibiotics on Grampositive bacteria. *Clinical Microbiology Reviews* 16: 175-88. doi:10.1128/CMR.16.2.175-188.2003.

Cantón, Emilia, Peman, Javier, Valentin, Amparo, Bosch, Maria, Espinel-Ingroff, Ana and Gobernado, Miguel. 2008. Comparison of posaconazole and voriconazole *in vitro* killing against *Candida krusei. Diagnostic Microbiology Infectious Disease* 62: 177-81. doi:10.1016/j.diagmicrobio. 2008.07.001.

Canton, Marilde and Onofre, Sideney B. 2010. Interference extracts of *Baccharis dracunculifolia* DC., Asteraceae, on the activity of antibiotics used in the clinical. *Brazilian Journal of Pharmacognosy* 20 (3): 348-54. doi.org/10.1590/S0102-695X2010000300010.

Carnelossi, Paulo R., Schwan-Estrada, Katia R. F., Cruz, M. E. S., Itako, A. T. and Mesquini, Renata M. 2009. Essential oils in the post-harvest control of *Colletotrichum gloeosporioides* in papaya. *Brazilian Journal of Medicinal Plants* 11(4): 399-406. doi.org/10.1590/S1516-05722009000400007.

Carson, Christine F., Hammer, Katherine A. and Riley, Thomas V. 2006. *Melaleuca alternifolia* (Tea Tree) oil: a Review of antimicrobial and on the medicinal properties. *Clinical Microbiology Reviews* 19(1): 50-62. doi: 10.1128/CMR.19.1.50-62.2006.

Chambers, Janice E. and Carr, Russel L. 1995. Biochemical mechanisms contributing to species differences in insecticidal toxicity. *Toxicology* 105: 291-304, 1995. doi.org/10.1016/0300-483X(95)03225-5.

Corazza, Sonia. 2004. Aromachology: a science of many smells. São Paulo: SENAC-São Paulo.

Correa-Royero, Julieth, Tangarife, Verónica, Durán, Camilo, Stashenko, Elena and Mesa-Arango, Ana. 2010. *In vitro* antifungal activity and cytotoxic effect of essential oils and extracts of medicinal and aromatic plants against *Candida krusei* and *Aspergillus fumigates*. *Brazilian Journal of Pharmacognosy* 20(5): 734-41. doi.org/10.1590/S0102-695X2010005000021.

Costa, Ane R. T., Amaral, Mônica F. Z. J., Martins, Paula M., Paula, Joelma A. M., Fiuza, Tatiana S., Tresvenzol, Leonice M. F., Paula, José R. and Bara, Maria Tereza F. 2011. Essential oil action *Syzygium aromaticum* (L.) Merr. and L. M. Perry on the hyphae of some pathogenic fungi. *Brazilian Journal of Medicinal Plants* 13: 240-45. Doi.org/10.1590/S1516-05722011000200018.

Dorman, Damien H. J. and Deans, Stanley G. 2000. Antimicrobial agents from plants: antibacterial activity of plant volatile oil. *Journal Applied Microbiology* 83: 308-16. Doi: 10.1046/j.1365-2672.2000.00969.x.

Duke, Stephen O., Romahni, Joanne G. and Dayan, Frank E. 2000. Natural products as sources for a new mechanism of herbicidal action. *Crop Protection* 19: 583-89. doi:10.1016/S0261-2194(00)00076-4.

Espinel-Ingroff Ana, Barchiesi, Francesca, Cuenca-Estrella, Manuel, Pfaller, Michael A., Rinaldi, Michael, Rodriguez-Tudela, Juan L. and Verweij, Paul E. 2005. International and multicenter comparison of EUCAST and CLSI M27-A2 broth microdilution methods for testing susceptibilities of *Candida* spp. to fuconazole, itraconazole, posaconazole, and voriconazole. *Journal of Clinical Microbiology* 43: 3884-889. doi:10.1128/JCM.43.8.3884-3889-2005.

Filippis, Flavia, M. 2001. Supercritical CO_2 extraction of essential oils Honsho and Ho-sho-experiments and modeling. 114f. Dissertation (Master´s in Chemical Engineering) – Department of Chemical Engineering, Federal University of Rio Grande do Sul.

Franz, Chlodwig M. 2010. Essential oil research: past, present and future. *Flavour Fragrance Journal* 25: 112-113. Doi: 10.1002/ffj.1983.

Gomes, Suellem P. and Favero, Silvio. 2011. Evaluation of essential oils of aromatic plants with insecticidal activity in *Triatoma infestans* (Klug, 1834) (Hemiptera: Reduviidae). *Acta Scientiarum* 33(2): 147-151. Doi: http://dx.doi.org./ 10.4025/actascihealthsci.v33i2.9531.

Haiduven, Donna. 2008. Nosocomial aspergillosis and building construction. *Medical Mycology* 25: 1-7. doi 10.1080/13693780802247694.

Hiremath, Geetha I., Ahn, Young J. and Kim, Soon I. 1997. Insecticidal activity of Indian plant extracts against Nilaparvata lugens (Homoptera; DelpHacidae). *Applied of Entomology Zoology* 32: 0152-166. doi.org/10. 1303/aez.32.159.

Isman, Murray B. 2006. Botanical insecticides, deterrents, and repellents in modern agriculture and an increasingly regulated world. *Annual Entomology* 51: 45-66. Doi: 10.1146/annurev.ento.51.110104.151146.

Jäger, Walter, Buchbaue, Gerhard, Jirovet, Leoplod and Fritzer, Monika. 1992. Percutaneous absorption of lavender oil from a massage oil. *Journal of the Society of Cosmetic Chemists* 43: 49-54. doi:10.1111/j.1468-2494.2010. 00579_2.x.

Jalali-Heravi, Mehdi, Zekavat, Berroz and Sereshti, Hassan. 2006. Characterization of essential oil components of Iranian geranium oil using gas chromatography–mass spectrometry combined with chemometric resolution techniques. *Journal of Chromatography* 1114: 154-16. doi:10. 1016/j.chroma.2006.02.034.

Knaak, Neiva and Fiuza, Lidia M. 2010. Potential of essentials oils from plants to control insects and microorganisms. *Neotropical Biology and Conservation* 5(2): 120-32. doi: 10.4013/nbc.2010.52.08.

Lavabre, Marcel. 2009. *"Aromatherapy: healing by essential oils; translation Raffaella de Filippis."* 7ª ed. Rio de Janeiro: Nova Era.

Lima, I. Oliveira, Oliveira, Rinalda A. G., Lima, Edeltrudes O., Farias, Nilma M. P. and Souza, Evandro L. 2006. Antifungal activity from essential oils on *Candida* species. *Brazilian Journal of Pharmacognosy* 16: 197-201. http://dx.doi.org/10.1590/S0102-695X2006000200011.

Lima, Maria P., Zoghbi, Maria das Graças B., Andrade, Eloisa Helena A., Silva, Tatiana M. D. and Fernandes, Carlos S. 2005. Volatile constituents of leaves and branches of *Cinnamonum zeylanicum* Blume (Lauraceae). *Acta Amazonica* 35: 363-66. Doi.1590/S0044-59672005003000009.

Lima, Rafaela K., Cardoso, Maria G., Santos, Custódio D., Moraes, Jair, C., Néri, Daniela K. P. and Nascimento, Evandro F. 2009. Chemical characterization of essential oil from leaves of guava (*Psidium guajava* l.) and its effects on the behaviour of cartepillar corn cartridge *Spodoptera frugiperda* (J. E. Smith, 1797) (Lepidoptera: Noctuidae). *Science and Agrotecnology* 33 (Special Edition): 1777-781. doi: http://dx. doi.org/10.1590/S1413-70542009000700013.

Machado, Bruna F. M. Teles and Fernandes Jr., Ary. 2011. Essential oils: general aspects and uses natural therapies. *Academic Notbooks* 3: 105-27. ISSN 2175-2532.

Maciel, Maria Aparecida M., Pinto, Angelo C., Veiga, Valdir E., Grynberg, Noema F. and Echevarria, Aurea. 2002. Medicinal Plants: The need for multidisciplinary studies. *New Chemistry*, 25: 429-38. doi.org/10.1590/ S0100-40422002000300016.

Matan, N., Rimkeeree, H., Mawson, A. J., Chompreeda, P., Haruthaithanasan, V. and Parker, M. 2006. Antimicrobial activity of cinnamon and clove oils under modified atmosphere conditions. *International Journal Food Microbiology* 107(2): 180-85. ISSN 0168-1605. Doi:10.1016/j. ijfoodmicro.2005.07.007.

Mendes, Sandra S., Bonfim, Rangel R., Jesus, Hugo C. R., Alves, Péricles B., Blank, Arie F., Stevan, Charles S., Antoniolli, Angelo R. and Thomazzi, Sara M. 2010. Evaluation of the analgesic and anti-inflammatory effects of the essential oil of *Lippia gracilis* leaves. *Journal of Ethnopharmacology* 129: 391-97. Accessed September 20, 2015. doi 10. 1016/j.jep.2010.04.005.

Militello, Marcello, Settanni, Luca, Aleo, Aurora, Mammina, Catarina, Moschettui, Giancarlo, Giammanco, Giovanni M., Blàzquez, Maria A. and Carrubba, Alessandra. 2011. Chemical composition and antibacterial potential of Artemisia arborescens L. essential oil. *Current Microbiology* 62: 1274-281. doi: 10.1007/s00284-010-9855-3.

Morais, Lilia A. S. 2009. Essential oils in phytosanitary control. In: Bettiol, W., Morandi, M. A. B. (Ed.) *Biocontrol of plant diseases: use and prospects*. Jaguariúna-SP: Embrapa Enviroment, Cap. 1, p. 341.

Mordue A. Jennifer (Luntz) and Nisbet, Alasdair J. 2000. Azadirachtin from de neem tree *Azadirachta indica*: its actions against insects. *Annais Sociedade Entomológica do Brasil* 29: 615-632. ISSN: 1809-8460. DOI: 10.1590/S0301-80592000000400001.

Muthukrishnan, J. and Pushpalatha, E. 2001. Effects of plant extracts on fecundity and fertility of mosquitoes. *Journal of Appied. Entomology* 125: 31-35. doi.org/10.1111/j.1439-0418.2001.00503.x.

Nascimento, Gislene G. F., Lucatelli, Juliana, Freitas, Paulo C. and Silva, Giuliana L. 2000. Antibacterial activity of plant extracts and phytochemicals on antibiotic-resistant bacteria. *Brazilian Journal Microbiology* 31: 247-56. doi.org/10.1590/S1517-83822000000400003.

Nguefack, Julienne, Budde, Birgitte B. and Jakobsen, Mogens. 2004. Five essential oils from aromatic plants of Cameroon: their antibacterial activity and ability to permeabilize the cytoplasmic membrane of *Listeria innocua* examined by flow cytometry. *Letters in Applied Microbiology* 39: 395-400.

Okoh, Omobola O., Sadimenko, Alexander P. and Afolayan, Abodunrin J. 2010. Comparative evaluation of the antibacterial activities of the essential oils of *Rosmarinus officinalis* L. obtained by hydrodistillation and solvent free microwave extraction methods. *Food Chemistry* 120(1): 308-312. doi:10.1016/j.foodchem.2009.09.084.

Oliveira, Maíra M. M., Brugnera, Danilo F., Cardoso, Maria G., Guimarães, Luiz Gustavo L. and Picolli, Roberta H. 2011. Yeld, chemical compositon and listerial activity of essential oils from species *Cymbopogon*. *Brazilian Journal of Medicinal Plants* 13(1): 8-16. doi. org/10.1590/S1516-05722011000100002.

Orafidiya, Lara O., Agbani, Ejaifi O., Iwalewa, Ezekiel O., Adelusola, Kayode A. and Oyedapo, Olaynka O. 2004. Studies on the acute and sub-chronic toxicity of the essential oil of *Ocimum gratissimum* L. leaf. *Phytomedicine* 11: 71-76. ISSN 0944-7113. Doi.10.1078/0944-7113-00317.

Organização Mundial de Saúde (WHO). 2010. "Financing health systems: the path to universal coverage". World Health Report. Geneva: World Health Organization. Doi.org/10.1590/S0102-311X2013000500003.

Oussalah, Mounia, Caillet, Stéphane, Saucier, Linda and Lacroix, Monique. 2007. Inhibitory effects of selected plant essential oils on the growth of four pathogenic bacteria: E. coli O157:H7, *Salmonella typhimurium*, *Staphylococcus aureus* and *Listeria monocytogenes*. *Food Control* 18(5): 414-20. doi.org/10.1016/j.foodcont.2005.11.009.

Pascholati, Sérgio F. and Leite, Breno. 1995. "Host: mechanisms of resistance." In: *Manual of Phytopatology – Principles and Concepts*, edited by Armando Bergamin Filho, Hiroshi kimati, Lilian Amorim, 417-54, Sao Paulo: Publishing Company Agronomic Ceres.

Pereira, Rogerio S., Sumita, Tânia C., Furlan, Marcos Roberto, Jorge, Antônio O. C. and Ueno, Mariko. 2004. Antibacterial activity of essential oils on microorganisms isolated from urinary tract infection. *Revista Saúde Pública* 38: 326-28. doi.org/10.1590/S0034-89102004000200025.

Pereira, Adriana C. R. L., Oliveira, José V., Gondim Jr., Manoel G. C. and Câmara, Claudio A. G. 2008. Insecticidal activity of essential oils and fixed on *Callosobruchus maculatus* (FABR., 1775) (Coleoptera: Bruchidae) in cowpea beans [*Vigna unguiculata* (L.)

WALP.]. *Science and Agrotecnology* 32(3): 717-24. doi: http://dx.doi.org/10.1590/S1413-70542008000300003.

Pinheiro, Antonio L. 2003. *Produção de óleos Essenciais*, Viçosa: CPT. p. 149.

Prates, Hélio T. and Santos, Jamilton P. 2002. "Essential oils in control of stored grain pests." In: *Armazenagem de grãos* edited by Ireneu Lorini, Lincon H. Miike and Vildes M. Scussel, 443-461, Campinas: Institute Bio Geneziz. Doi.org/10.1059/S1413-70542008000300003.

Rajkumar, Sankaran and Jebanesan, Arulsamy. 2010. Chemical composition and larvicidal activity of leaf essential oil from *Clausena dentata* (Willd) M. Roam. (Rutaceae) against the chikungunya vector, *Aedes aegypti* Linn. (Diptera: Culicidae). *Journal of Asia-Pacific Entomology* 13: 107-109. doi:10.1016/j.aspen.2010.02.001.

Runyoro, Debora, Ngassapa, Olipa, Vagionas, Konstantinos, Aligiannis, Nektarios, Graikou, Konstantia and Chinou, Ioanna. 2010. Chemical composition and antimicrobial activity of the essential oils of four Ocimum species growing in Tanzania. *Food Chemistry* 119: 311-16. doi. org/10.1016/j.foodchem.2009.06.028.

Sabulal, Baby, Dan, Mathew J. A. J., Kurup, Rajani, Pradeep, Nediyamparambu S. and Valsamma, Renju K. and George, Varughese. 2006. Caruophyllene-rich rhizome oil of *Zingiber nimmonni* from South India: Chemical characterization and antimicrobial activity. *Phytochemistry*, 67: 2469-473. doi:10.1016/j.phytochem.2006.08.003.

Santos, Gil R., Brum, Rúbia B. C. S., Castro, Henrique G., Gonçalves, Clebson G. and Fidelis, Rodrigo R. 2013. Effect of essential oils of medicinal plants on leaf blotch in Tanzania grass. *Revista Ciência Agronômica* 44(3): 587-93. doi.org/10.1590/S1806-66902013000300022.

Santurio, Jânio M., Santurio, Deise F., Pozzatti, Patricia, Moraes, Cristiane, Franchin, Paulo R. and Alves, Sydney H. 2007. Antimicrobial activity of the essential oils of oregano, thyme and cinnamon against serovars *Salmonella enterica* avicola of origin. *Rural Science* 37(3): 803-08. doi. org/10.1590/S0103-84782007000300031.

Santurio, Deise F., Costa, Mateus M., Maboni, Grazieli, Cavalheiro, Carlos P., Sá, Mariangela F., Pozzo, Marcelo D., Alves, Sydney H. and Fries, Leadir L. M. 2011. Antimicrobial activity of essential oils of spice against samples *Escherichia coli* isolated from poultry and cattle. Rural Science 41(6): 1051-56. doi.org/10.1590/s0103-84782012000800023.

Sartoratto, Adilson, Machado, Ana Lucia, M., Delarmelina, Camila, Figueira, Glyn M., Duarte, Marta C. T. and Redher and Vera Ligia G. 2004. Composition and antimicrobial activity of essential oils from aromatic plants used in Brazil. *Brazilian Journal Microbiology* 35: 275-280. http:// dx.doi.org/10.1590/S1517-83822004000300001.

Schelz, Zsuzsanna M. J. and Hohmann, Judit. 2006. Antimicrobial and antiplasmid activities of essential oils. *Fitoterapia* 77: 279-85. doi:10. 1016/j.fitote.2006.03.013.

Silva, Saulo L., Chaar, Jamal S., Figueiredo, Patricia M. S. and Yano, T. 2008. Cytotoxic evaluation of essential oil from *Casearia sylvestris* Sw on human cancer cells and erythrocytes. *Acta Amazônica* 38: 107-12. doi.org/ 10.1590/S0044-59672008000100012.

Silva, Mariama T. N., Ushimaru, Priscila I., Barbosa, Lidiane N., Cunha, Maria de Lourdes R. S. and Junior, Ary F. 2009. Antibacterial activity of plant essential oils against *Staphylococcus aureus* and *Escherichia coli* strains isolated from human specimens. *Brazilian Journal Medicinal Plants* 11: 257-62. doi.org/10.1590/S1516-05722009000300005.

Silva, Mayara G. F. 2011. Antioxidant and antimicrobial *in vitro* activity of essential oils and hydroalcoholic of marjoram (*Origanum majorana* L.) and basil (*Ocimum basilicum* L.). 2011. p. 70. Work of Course Completion, Pato Branco-PR – Federal Technological University of Parana.

Silveira, Sheila M., Cunha Jr. and Anildo, Scheuermann, Gerson N., Secchi, Fabio L. 2012. Chemical composition and antimicrobial activity of essential oils from selected herbs cultivated in the South of Brazil against food spoilage and foodborne pathogens. *Ciência Rural* 42: 1300-1306. doi.org/10.1590/S0103-84782012000700026.

Silver, Lynn L. and Bostian, Keith A. 1993. Discovery and development of new antibiotics: the problem of antibiotic resistance. *Antimicrobial Agents and Chemotherapy* 37: 377-83. doi:10.1128/AAC.37.3.377.

Simões, Claudia Maria O., Schenkel, Eloir P., Gosmam, Grace, Mello, João Carlos P., Mentz, Lilian A. and Petrovick, Pedro R. 2003. *Farmacognosia da Planta ao medicamento.* 5ª ed., Porto Alegre, Universidade, p. 1102. doi.org/10.1590/S0102-695X2002000100005.

Snelders, Eveline, Van Der Lee, Henrich A., Kuijpers, Judith, Rijs, Antonius J., Varga, Janos, Samson, Robert A., Mellado, Emilia, Donders, Rogier A., Melchers, Willem J. and Verweij, Paul E. 2008. Emergence of azole resistance in *Aspergillus fumigatus* and spread of a single resistance mechanism. *PLoS Med.* 11: 1629-637. doi: 10.1371/journal.pmed. 0050219.

Soares, Cristina S. A., Silva, Marise, Costa, Marlice B. and Bezerra, Carlos E. S. 2011. Insecticidal action of essential oils on the defoliating *Thyrinteina arnobia* (Stoll) (Lepidoptera: Geometridae). *Green Journal* 6 (2): 154-57. Doi: http://dx.doi.org/10.18378/rvads.v10i3.

Soares, Cristina S. A., Silva, Marise, Costa, Marlice B., Bezerra, Carlos E. S., Carvalho, Livia M. and Soares, Alexandre H. V. 2012. Inseticidal activity of essential oils on *Macrosiphum euphorbiae* (Thomas) (Hemiptera: Aphididae) in rosebush. *Brazilian Journal of Agroecology* 7 (1): 169-75. ISSN: 1980-9735.

Soylu, Emine M., Kurt, Sener and Soylu, Soner. 2010. *In vitro* and *in vivo* antifungal activities of the essential oils of various plants against tomato grey mould disease agent *Botrytis cinerea*. *International Journal of Food Microbiology* 143(3): 183-189. doi: 10.1016/j.ijfoodmicro.2010.08.015.

Steffani, Evandro. 2003. Mathematical modeling of the supercritical extraction process of essential oil Ho-Sho (*Cinnamomum camphora* Nees and Eberm var. linaloolífera Fujita) Utilizando CO_2. Thesis (Doctoral) – Federal University of Santa Catarina.

Steffens, Andreia H. 2010. "Study of the chemical composition of the essential oil obtained by steam distillation at laboratory scale and industrial." Dissertation, Pontifical Catholic University Catolica of Rio Grande do Sul.

Stevensen, C. Joseph. 1998. Aromatherapy in dermatology. *Clinics in Dermatology* 16: 689-94.

Strapazzon, Julia Ornelas. 2004. "Chemical composition and antimicrobial analysis of volatile oil *Annona squamosa* L. (Ariticum.) Chapeco, SC." Graduate Thesis. University Comunnity Regional Chapeco.

Tanida, Mamoru, Niijima, Akira, Shen, Jiao, Nakamura, Takuo and Nagai, Katsuya. 2005. Olfactory stimulation with scent of essential oil of grapefruit affects autonomic neurotransmission and blood pressure. *Brain Research* 1058(1-2): 44-55. ISSN 0006-8993.

Ultee, Annemieke, Bennik, Marjon H. and Moezelaar, Roy. 2002. The phenolic hydroxyl group of carvacrol is essential for action against the food-borne pathogen *Bacillus cereus. Applied Environmental Microbiology* 68: 1561-568. doi: 10.1128/AEM.68.4.1561-1568. 2002.

Viegas Júnior, Cláudio. 2003. Terpenes with insecticidal activity: an alternative to the chemical controlo f insects. New Chemical 26: 390-400. doi.org/10.1590/S0100-40422003000300017.

Vieira, Gustavo H. C., Andrade, Wagner P. and Nascimento, Daniela M. 2012. Use of essential oils in control of *Varroa destructor* in *Apis mellifera. Search Tropical Farming* 42(3): 317-322. doi: 10.5216/ Pat.

Wannes, Wissen A., Mhandi, Baya, Siriti, Jazia, Jemia, Marrien B., Ouchich, Olfa, Hamdaoui, Ghaith, Kchouk, Mohamed E. and Marzouk, B. 2010. Antioxidant activities of the essential oils and methanol extracts from myrtle (*Myrtus communis* var. italica L.) leaf, stem and flower. *Food and Chemical Toxicology* 48: 1362-1370. Accessed September 19, 2015. doi: 10.1016/j.fct.2010.03.002.

Wheeler, Deborah A. and Isman, Murray B. 2001. Antifeedant and toxic activity of *Trichilia americana* extract against the larva of *Spodoptera litura. Entomology Experimental Applied* 98: 16, 2001. doi.org/10.1046/j. 1570-7458.2001.00751.x.

Worwood, Susan. 1995. *Aromatherapy. A guide from A to Z for the therapeutic use of essentials oils.* Sao Paulo, Publishing Company BestSeller.

Zago, Juliana A. A., Ushimaru, Priscila I., Barbosa Lidiane N. and Fernandes Jr. Ary. 2009. Synergism between essential oils and antimicrobial drugs on strains of *Staphylococcus aureus* and *Escherichia coli* isolated from clinical cases human. *Brazilian Journal of Pharmacognosy* 19(4): 828-33. doi.org/10.1590/S0102-695X2009000600005.

Zhao, Boguang, Grant, G. G., Langevin, D. and MacDonald, L. 1998. Deterring and inhibiting effects of quinolizidine alkaloids on spruce budworm (Lepidoptera: Tortricidae) oviposition. *Environmental Entomology* 27: 984-92. Doi: http://dx.doi.org/10.1093/ee/27.4.984 984-992.

In: Essential Oils
Editor: Miranda Peters
ISBN: 978-1-63484-351-5

Chapter 2

PHENOLOGICAL CHANGES IN THE BIOSYNTHESIS AND CHEMICAL COMPOSITION OF THE ESSENTIAL OILS

Jaime Usano-Alemany[1,*], Jesús Palá-Paúl[2] and David Herráiz-Peñalver[3]
[1]Institute of Crop Science and Resources Conservation, INRES-Horticultural Sciences, University of Bonn, Bonn, Germany
[2]Department of Plant Biology I (Botany and Plant Physiology), Faculty of Biology, Complutense University, Madrid, Spain
[3]Centre for Research in Agroforestry Albaladejito (CIAF-IRIAF), Junta de Comunidades de Castilla-La Mancha, Cuenca, Spain

ABSTRACT

The biosynthesis of the essential oils can be affected by a number of factors, mainly genetic and environmental ones. These potential variations in their chemical composition can occur and have been the subject of many researchers. Since many of the most commercially important species are perennial species, their harvest time and cultivation conditions can be upgraded through better knowledge of the variations in the phenology in order to improve yields and quality of their essential oil. Furthermore, examining the trend of qualities and yields over 5, 6 or even more years throughout the plant development provides a key tool for the strategic establishment of new promising species as alternative crops. Producers are interested in these alternative crops that may improve the environmental and economic sustainability of existing cropping systems. In most species, the chemical composition of the essential oils is mainly determined by individual variations and, therefore, is largely subjected to a genetic control. However, ontogeny changes throughout phenology can cause more or less pronounced variations in a few compounds of the whole essential oil chemical spectra. Notwithstanding these slight variations in chemical composition, the phenological stages have a greater influence on the yields obtained from distillation of dry material. The most recent published data leads

* Corresponding author: jusanoal@uni-bonn.de; jaimeusano@yahoo.es.

to the suggestion that although flowering time is the most common harvest time for essential oil, producing plants may not always be in line with the highest yields. Herein, we provide an overview of the most relevant essential oil changes regarding plant phenology and plant development, summarize the latest reports related to the monitoring of essential oil production, and outline their implications for productivity and quality.

INTRODUCTION

Today there continue to be recently published reports about novel and upgraded applications and properties of diverse plant extracts (Lubbe and Velpoorte 2011). Among them, essential oils have a longstanding reputation as natural products with medicinal properties (Kennedy and Wightman 2011; Raut and Karuppayil 2014), antibacterial and antioxidant activities (Amorati et al., 2013; Teixeira et al., 2013), antifeedant and biopesticidal effects (Badawy et al., 2010; Kumar et al., 2011; Isman et al., 2011; González-Coloma et al., 2011), natural food preservatives properties (Hyldgaard et al., 2012; Calo et al., 2015; Prakash et al., 2015), veterinary effects (Ellse and Wall 2014), among others reported. This indicates an increasing demand for essential oils in many areas. However, the efficacy and safety of these bio-products is sometimes diluted due to well reported lack of reproducibility. This could be because essential oils are very complex mixtures of low molecular weight terpenoids and phenylpropanoids as main constituents with different degrees of volatilities and polarities which are subjected to an inherent biological variation.

Sources of Variability

The above mentioned biological variations in the secondary metabolite products may have a genetic origin causing not only differentiation between plant species of the same botanical family or genus (Janicsák et al., 2006; Rzepa et al., 2009; Adams, 2010), but also intra-specific variability giving rise to different chemotypes within a single species (Böszörményi et al., 2009; Gnavi et al., 2010; Shojaiefar et al., 2015). Moreover, there are many external factors which can influence the biosynthesis of essential oils in plants as environmental and agronomic conditions (Ormeño et al., 2009, Tibaldi et al., 2011, Yosr et al., 2013), stress factors (Bettaieb et al., 2009; Selmar and Kleinwächter 2013b), time of harvest and phenology (Jordán et al., 2006, Moghaddam et al., 2015; Smitha and Virendra 2015), different parts of the plants used for the extraction (Bettaieb et al., 2010, Ben Marzoug et al., 2011) and even epigenetic changes resulting from concrete growing conditions (Lavania et al., 2012; Ohlsson et al., 2013). Besides, the method of extraction and storage are other sources of variability related to the product management which may vary the physico-chemical properties of the essential oils (Silva et al., 2014; Mendez-Tovar et al., 2015). Figueiredo et al. (2008) and Barra (2009) showed a list of these major sources of variation in the essential oils. Considering that properties, effectiveness and therapeutic potential of a given species are the result of a concrete chemical composition and possible synergy of available chemical entities, this wide range of sources of variability brings regularly changes in the assumed efficacy of the essential oils and other plant natural products. Consequently, using the binomial nomenclature of the species in manuscripts reporting biological properties of diverse plant extracts or describing phytochemical profiles of a species from a single or few samples might not be enough to link the putative pharmacological effectiveness or other biological properties with a well-defined herbal product (Rivera et al., 2014).

There are many examples of this chemical variability of essential oils and implications for the final quality of the product. *Salvia lavandulifolia* is a native Mediterranean shrub from the Iberian Peninsula and North-West Africa which has a longstanding and well reported reputation as a plant with essential oil with remarkable pharmacological properties. It is known for having 1,8-cineole as a major compound but in comparative studies of wild populations also high content of other monoterpenes such as camphor, limonene, α-pinene, β-pinene and even sesquiterpenes such as viridiflorol and spathulenol can be achieved (Herráiz-Peñalver et al., 2010; Usano-Alemany et al., 2014a). In this case, 1,8-cineole has been shown to be the most potent single component in terms of cholinesterase inhibition which may have serious implications for treatments against brain diseases such as Alzheimer or dementia (Savelev et al., 2003; Kennedy and Wightman 2011; Miroddi et al., 2014). Nonetheless, its content within the essential oils of *S. lavandulifolia* change significantly to the extent that it has been reported a range of variation between 8.2% and 75% (Usano-Alemany and Herráiz 2014), and hence the expected bioactivity for this product may has a poor accuracy, repeatability and reproducibility. Furthermore, this compound showed also a large variability when reported in many other species (Table 1). In fact, among the top essential oils commercialized according to the global production (Lubbe and Velpoorte 2011), numerous species show large differences between essential oil major components, and hence the quality and properties in most cases are restricted by the natural variability. That is the case for *Eucalyptus* species in which essential oil chemical composition varies between species and seasons (Ben et al., 2012). Changes in the amount of the active compounds determine their fumigant toxicity against pest insects as a relevant biological property. Another example that emerged is the spike lavender (*Lavandula latifolia*) whose essential oils have as major components 1,8-cineole, camphor and linalool, accounting for more than 70% in most samples. In this case, Herráiz-Peñalver et al. (2013) described certain relationship with the geographical origin of the samples and the frequency distributions of the major compounds in their relative percentages very close in shape to a normal Gaussian distribution unimodal type. The phenomenon of normal Gaussian distribution of essential oil components is in agreement with the statement that a substantial number of additional and independently acting causes determine the biosynthesis of essential oils and other secondary metabolites. The variability of the essential oils depends on internal and external factors and plays a key role in the chemical composition. In order to encompass the expected variability in the internal quality components, International Standards (ISO) provide percentage value ranges for the most representative compounds. These standards are reviewed regularly to ensure market quality requirements. Among the most recent and updated ISO norms for essential oils are included: ISO 16928:2014 [Essential oil of ginger (*Zingiber officinale* Roscoe)], ISO 9841:2013 [Essential oil of hyssop (*Hyssopus officinalis* L. spp. *officinalis*)], ISO 1342:2012 [Essential oil of rosemary (*Rosmarinus officinalis*)], ISO 3526:2012 [Essential oil of Spanish sage (*Salvia lavandulifolia*)], among many others.

Table 1. Variation in the content of 1,8-cineole in the essential oil

Species	Common name	Content*	References
Artemisia pontica	Roman wormwood	12-23	De Vicenci et al., 2002
Eucalyptus globulus	Tasmanian blue gum	70-80	De Vicenci et al., 2002
Eucalytus camaldulensis	River red gum	25-44	Tsiri et al., 2003
Eucalyptus polybractea	Blue Mallee	>70	Goodger and Woodrow, 2009
Rosmarinus officinalis	Rosemary	12-47	De Vicenci et al., 2002
Salvia officinalis	Sage	8-23	De Vicenci et al., 2002
Salvia lavandulifolia	Spanish sage	12-41 6-59 1-54 8-75	De Vicenci et al., 2002 Schmiderer et al., 2008 Kintzios, 2000 Usano-Alemany et al., 2014
Elettaria cardamomum	Green cardamom	13-51	De Vicenci et al., 2002
Hedichium flavum	Yellow ginger	42	De Vicenci et al., 2002
Mentha piperita	Peppermint	15-18	De Vicenci et al., 2002
Thymus vulgaris [chemotype 1,8-cineol]	Common thyme	36	Jordán et al., 2006
Thymus albicans	-	50-66	Miguel et al., 2004
Lavandula latifolia	Spike lavender	28-35 6-57	Salido et al., 2004 Herráiz-Peñalver et al., 2013

*All data presented are referred to mean values published as %.

The implications of these sources of variability remain in most cases unresolved and only in species in which previous long-term work programmes of breeding, selection and cultivation have been conducted or are nowadays in progress e.g., *Salvia officinalis* (Böszörményi et al., 2009; Mader et al., 2010; Grausgruber-Gröger et al., 2012), *Salvia lavandulifolia* (Herráiz-Peñalver et al., 2010; Usano-Alemany et al., 2014a,b) *Rosmarinus officinalis* (Jordán et al., 2012, 2013, 2014), *Mentha x piperita* (Dorman et al., 2003; Grulova et al., 2015) or *Eucalyptus* sp. (Eldridge et al., 1993; Elaissi et al., 2011; Andrew et al., 2013), to name but a few, are beginning to become known. Nonetheless, in other species with highly promising application through their diverse plant extracts (Mukherjee et al., 2011; Vanzani et al., 2011; Howes and Hougton 2012; Gechev et al., 2014) their genetic and biochemical variability along with their cultivation requirements remain in most cases unknown. This can have major consequences on repeatability, resulting in struggling to a standardization of plant natural products. Furthermore, over the last decade, molecular engineering has enabled us to gain a better understanding of the regulation of the methabolic pathways which lead to the biosynthesis of essential oils. Besides, transgenic aromatic plants are enabled to accumulate elevated levels of specific terpenoids by the overexpresion of a particular terpene synthase gene (Lange and Ahkami 2013; Mendoza-Pouderaux et al., 2014).

Table 2. Essential oil yield data reported from different phenological stages in representative wild and cultivated aromatic and medicinal plants

Species	Plant material		Phenology*				References
	Cultivated	Wild	Vegetative	Floral budding	Flowering	Fruiting	
1. *Hypericum perforatum* St John's-wort	-	•	0.07	0.08	0.09	0.06	Schwob et al. (2004) **(1.)**
2. *Hypericum triquetrifolium* Curled-leaved St. John's-wort	-	•	0.09	-	0.12	0.08	Hosni et al. (2011) **(2.)**
3. *Rosmarinus officinalis* Rosemary	- - •	• • -	- 0.64 0.48	- - -	1.55 0.71 -	2.45 0.40 0.2	Jordán et al. (2013) **(3.1)** Yosr et al. (2013) **(3.2)** Lakušić et al. (2013) **(3.4)**
4. *Thymus caramanicus*	-	•	1.90	2.10	2.50	2.00	Nejad Ebrahimi et al. (2008) **(4.)**
5. *Thymus algeriensis*	-	•	2.40	-	1.80	-	Zouari et al. (2012) **(5.)**
6. *Thymus hyemalis* Winter thyme	•	-	3.13	-	3.52	2.63	Jordán et al. (2006) **(6.)**
7. *Thymus vulgaris* Common thyme	• •	- -	4.73	- 2.08	3.99 1.80	2.33 1.82	Jordán et al. (2006) **(7.1)** Naghdi Badi et al. (2004) **(7.2)**
8. *Thymus maroccanus* Moroccan wild thyme	-	•	2.14	-	2.73	1.72	Jamali et al. (2013) **(8.)**
9. *Pelargonium graveolens* Rose geranium	•	-	0.16	0.14	0.18	0.16	Boukhris et al. (2013) **(9.)**
10. *Salvia lavandulifolia* Spanish sage	• • •	• - -	0.88 1.54 1.50	1.02 - -	1.26 1.75 2.00	1.95 - 3.10 -	Usano-Alemany et al. (2014) **(10.1)** Porres-Martínez et al. (2014) **(10.2)** Herráiz-Peñalver et al. (2015) **(10.3)**

Table 2. (Continued)

Species	Plant material		Phenology*				References
	Cultivated	Wild	Vegetative	Floral budding	Flowering	Fruiting	
11. *Salvia officinalis* Sage	• • •	- - -	0.72 0.7 0.6	1.00 0.9 -	1.40 0.5 0.9	0.92 0.3 1.0	Arraiza et al. (2012) **(11.1)** Mirjalili et al. (2006) **(11.2)** Herráiz-Peñalver et al. (2015) **(11.3)**
12. *Ocimum ciliatum*	•	-	1.30	1.80	1.16	0.91	Moghaddam et al. (2015) **(12.)**
13. *Lavandula latifolia* Spike lavender	-	•	-	-	1.85	0.45	Salido et al. (2004) **(13.)**
14. *Lavandula angustifolia* English lavender	•	-	0.5	-	0.6	8.1	Lakušić et al. (2014) **(14.)**
15. *Cymbopogon martini* Gingergras	•	-	-	1.53	1.46	1.97	Smitha and Virendra et al. (2015) **(15.)**
16. *Origanum vulgare* Oregano, Wild marjoram	• -	- •	84.51 dm^3ha^{-1} 1.85	215.51 dm^3ha^{-1} -	325.45 dm^3ha^{-1} 0.60	181.04 dm^3ha^{-1} -	Barakauskiené et al. (2013) **(16.1)** Béjaoui et al. (2013) **(16.2)**
17. *Origanum majorana* Sweet marjoram	•	-	0.05	0.07	0.09	-	Sellami et al. (2009) **(17.2)**
18. *Lippia citriodora* Lemon verbena	•	-	0.48	-	0.90	0.25	Shahhoseini et al. (2013) **(18.1)**
19. *Hyssopus officinalis* Hyssop	-	•	0.04	0.14	0.40	0.24	Khan et al. (2012) **(19.)**
20. *Satureja montana* Winter savory	- •	• -	1.73 2.15		1.56 1.20	0.70 2.20	Milos et al. (2001) **(20.1)** Mirjana et al. (2004) **(20.2)**
21. *Artemisia scoparia* Redstem wormwood	-	•	0.40	0.70	0.90	-	Mirjalili et al. (2007) **(21.)**

* All data presented are referred to mean values published as % (v/w). Data from percentages of the main essential oil components is presented in Table 4 according to reference numbers.

Ecological Role of Volatile Organic Compounds

The release of volatile organic compounds (VOCs) provides plants a way to communicate with organisms across distances. These VOCs are released from leaves, flowers and fruits into the atmosphere and from roots into the soil. However, the study of the physiology and function of these volatile organic compounds released into the environment further than attracting pollinators by flowers, is still partly unknown and subject to intensive research (Pichersky and Gershenzon 2002; Gershenzon and Dudareva 2007; Maffei et al., 2011). Many species store their volatile organic compounds i.e., essential oils, in specialized glandular trichomes which release their contents in response to tissue damage. Research on volatile emissions by plants shows that VOCs are potent signaling molecules that have evolved to serve multiple functions (Mafei 2010). The major identified function of terpenoids (monoterpenes, sesquiterpenes and diterpenes) is related to their defensive properties. Plants as sessile organisms require defensive strategic weapon systems against predators, diseases or any kind of stress situations. Essential oils play a very important role in the defensive strategies and this fact leads to potential uses in the protection of crops against pests (Isman et al., 2011). Due to different environmental situations that plants must face during their biological cycle, the biosynthesis of essential oils and hence the presence of glandular trichomes density, morphology and chemistry may undergo a remarkable variation throughout the different developmental stages (Schmiderer et al., 2008) and under multiple kind of biotic and abiotic stress situations (Karray-Bouraoui et al., 2009; Harada et al., 2010; Yadav et al., 2014). As a result, changes with phenology in yield and chemical composition of the essential oils have been noticed in many species. The work on understanding this source of variability will contribute significantly to a better management of well-established aromatic and medicinal plants crops and will play a decisive role in helping promising species to consolidate as alternative crops.

Phenological Changes in the Biosynthesis of Essential Oils

Secretory Structures

Glandular trichomes are specialized hairs found on the surface of about 30% of all vascular plants and are responsible for a significant portion of a plant´s secondary chemistry (Glas et al., 2012). The density of glandular trichomes on developing leaves can be estimated using scanning electron microscopy (Figure 1). Typically, young leaves contained fewer glandular trichomes than older leaves indicating an evident gland production during leaf growth (Maffei, 2010). The development of all secretory structures, which is asynchronous, occurs during leaf expansion and differentiation (De Luna et al., 2014). In many plant species new glandular trichomes are continually produced during leaf growth and newly initiated glands do occur together with mature glands in growing regions, such that neighboring glands within the same leaf zone are often of different ages. This fact provides even gland to gland variability depending on the region of the leaf (Baran et al., 2010; Boukhris et al., 2013), although differences within young leaves are not as pronounced as in older leaves. Obviously, a strong correlation is given between the formation of secretory structures and the essential oil biosynthesis, so different maturation stages have shown remarkable variations in the essential oil yield. Nonetheless, there is some controversy regarding variations on yields at different phenological stages. It is not unusual to find plenty of references with different outcomes in this regards, sometimes even with conflicting data. Results largely depend on firstly the species which is described and then the length of the experiments, the environmental and/or culture conditions, the number of harvest times per year and eventually the phytochemical biodiversity within the plant material. Thus, maximum yields throughout the plant development in many species have been reported for beginning or full bloom and fruit-seed maturation and, to a lesser extent for the vegetative stage (Table 2). For example, highest essential oil yields at the flowering time stage were obtained by Yosr et al. (2013) for *Rosmarinus officinalis* which in turn are in agreement with those obtained earlier by Hamrouni Sellami et al. (2009), Rohloff et al. (2005) and Naghdi Badi et al. (2004) for *Origanum majorana*, *Mentha* x *piperita* and *Thymus vulgaris*, respectively. The highest amount of essential oil was obtained from flowering plants in all cases although with remarkable differences in both the number of analyzed samples and the term of monitoring. However, in other cases, maximum yield has been reported when plants begin with fruit maturation even though in some species with conflicting reports. This is particularly the case for *Rosmarinus officinalis* in which Jordán et al. (2013) and Singh and Guleria (2013) found that the seed setting stage produced the most essential oil. Other examples of major yields when seed settings are those reported by Usano-Alemany et al. (2014a) or Smitha and Virendra (2015) in *Salvia lavandulifolia* and *Cymbopogon martinii*, respectively. In this connection, it has been suggested that higher concentrations of phenolic compounds (include phenolic acids, flavonoids, tannins, and others) can be found during flowering whilst terpene concentrations increased during fructifications (Abreu et al., 2004). Only in a limited number of cases vegetative phases showed as the dominant phenological stage regarding essential oil yield. For example, Jordán et al. (2006) and Béjaoui et al. (2013) affirmed this for *Thymus vulgaris* and *Origanum vulgare*, respectively. Nevertheless, the authors themselves described the beginning of fruit maturation as prominent in *Thymus hyemalis* (Jordán et al., 2006). On the other hand, Mirjalili et al. (2006) collected more oil in the vegetative stage than flowering and fruiting in *Salvia officinalis*. Furthermore, in other cases variations in essential oil content were asynchronous with different maximum and minimum values throughout the year in cultivated *S. officinali*s in two different sites (Santos-Gomez and Fernandes 2001).

The Importance of the Plant Organ Used for the Distillation

Differences in essential oil yield according with phenology can be at least partially explained in most cases by remarkable variability recorded from the different plant parts used for the extraction. In this regard, each plant species present certain organs in which essential oils are accumulated in higher amounts (Table 3). This has been reported when different plant parts have been distilled separately. Distilled flowers or inflorescences showed the highest essential oil yields in species as diverse as *Eryngium planum* and *E. aquifolium* (Palá-Paúl et al., 2010; Thiem et al., 2011), *Michelia compressa* (Su et al., 2015), *Salvia officinalis* (Perry et al., 1999), *Cominum cyminum* (Bettaieb et al., 2009) or *Lavandula angustifolia* (data not published) while in other species as *Eucalyptus oleosa* (Ben Marzaug et al., 2011) or *Ruta chalepensis* (Tounsi et al., 2011) fruits showed maximum essential oil concentrations. In contrast, Spanish *Eryngium glaciale* showed a higher essential oil content in roots, followed by the stems and leaves and finally the inflorescence (Palá-Paúl et al., 2005). This spatial accumulation of essential oil in the different plant parts often results in increases and decreases in yield during phenology, largely explaining this way the seasonal dynamic accumulations. As different relationships between biomass in stems, leaves, flowers and/or fruits can be obtained throughout phenology when plants are harvested, this fact may have remarkable implications in the final essential oil yield achieved. In some cases such as roses, citruses or even in lavender plants (where only bloomed spikelets along with few leaves and stems are used) the isolation of essential oils is undertaken by distillation-extraction of concrete parts and hence yield is given solely by the content of concrete organs. Nonetheless, species in which whole aerial parts are subjected to distillation such as sage species, thymus, or rosemary, to name but a few, the relation between leaves, stems and flowers is an important fact to consider. In these cases, yields in the flowering stage could not be estimated with accuracy if only results from flowers were taken into consideration.

Table 3. Essential oil yield data reported from organs in representative wild and cultivated aromatic and medicinal plants

Species	Plant material		Plant part*				References
	Cultivated	Wild	Leaves	Steams	Flowers	Fruits	
1. *Ruta chalepensis* Fringed rue	•	•	0.48	0.57	0.84	2.10	Tounsi et al. (2011)
2. *Thymus maroccanus* Moroccan wild thyme	-	•	1.64	-	3.46	2.46	Jamali et al. (2013)
3. *Rosmarinus officinalis* Rosemary	-	•	1.17	<0.01	0.05	-	Yosr et al. (2013)
4. *Psorolea bituminosa* Bitumen trefoil	-	-	0.10	-	0.20	0.20	Bertoli et al. (2004)
5. *Cominum cyminum* Cumin	•	-	0.1	0.1	1.7	-	Bettaieb et al. (2010)
6. *Hyssopus officinalis* Hyssop	-	•	4.20	0.22	4.40	-	Pandey et al. (2014)
7. *Porcelia macrocarpa* Monkey banana	•	-	0.05	-	-	0.06	Da Silva et al. (2013)
8. *Eucalyptus oleosa* Red mallee	•	-	0.45	0.52	0.52	1.12	Ben Marzoug et al. (2011)
9. *Salvia officinalis* Sage	•	-	1.11	0.05	1.56	-	Perry et al. (1999)
10. *Eryngium planum* Blue eryngo (roots with 0.05%)	-	•	0.07	0.10	0.23	-	Thiem et al. (2011)

* All data presented are referred to mean values published as % (v/w).

Data from percentages of the main essential oil components is presented in Table 6 according to species numbers.

The Role of Stress

Another source of variation according to the phenology is the level of stress due to the environmental conditions. Hence, plants may boot the production of natural compounds in response to stress conditions (Selmar and Kleinwächter 2013a). Stress may come from pathogens, herbivores, parasites or plant-plant competence which is classified as biotic stress and also from damage due to dryness or flooding, light and radiation, temperatures, salinity, nutrient deficit or metal toxicity in which case it is called abiotic stress. Plants face different level of stress throughout the phenological cycle and this may have remarkable implications on the essential oil yield. For instance, the drought stress (both natural and induced) showed higher accumulation of secondary metabolites, essential oils among them (Selmar and Kleinwächter 2013a,b). Accordingly, biosynthesis of essential oil in numerous species grown in Mediterrean or semi-arid regions seems to be higher in summer time, which is in line in numerous cases with the flowering time. On the other hand, enhancement of essential oil accumulation in flowers and fruits and in some cases in leaves during phenology may have a biological and ecological significance. It is noteworthy that floral scent plays numerous important roles in the interaction between plants and their surrounding pollinators, phytophageous, etc. In other cases, an increase in output of essential oil during seed formation and maturation (particularly in autumn time) may be related to an increase in the protection against pathogens and fungi (González-Coloma et al., 2011; Teixeira et al., 2013; Saviuc et al., 2015) and to avoid growth of seeding which can cause competence (Sánchez-Muñoz et al., 2012; Santana et al., 2012).

Genetic vs Environmental Modulation of the Essential Oil Production

In crops consisting of heterogeneous plants from a single species, potential individual variation in yields among plants should be considered. Similarly, significant variations among and within population have been reported in wild collected individual plants (Homer et al., 2000; Herráiz-Peñalver et al., 2013; Rajabi et al., 2014). When wild populations are in fact looked at in different geographical and hence environmental conditions, yields often vary substantially in relation to, for instance, altitude levels as reported by Khadivi-Khub et al. (2014) for *Stachys lavandulifolia*, being also possible to discern between individuals with top yields as done by Muñoz-Bertomeu et al. (2007) for *Lavandula latifolia*. Nevertheless, if concrete genotypes from different geographic origins are cultivated in the same plot, phenotypic yield variances among individuals are difficult to detect after years of monitoring (Usano-Alemany et al., 2014a; Shojaiefar et al., 2015). In these studies, whilst oil yield fluctuates in response to different environmental conditions and phenology, essential oil composition broadly remains constant for each individual. This means that selection for high essential oil yield accessions may need more detailed survey and long-term monitoring than selection for superior essential oil chemotypes which seems to be more genetically controlled. Thus, in relation to the phenological changes in the essential oil yield, intra-specific biodiversity has been reported, but this is highly influenced by environmental conditions and/or stress factors. Recent reports in several organisms support that metabolic flux through terpene pathways is mainly controlled at enzymatic level regulated transcriptionally and post-transcriptionally in response to metabolic, environmental and developmental cues (Córdoba et al., 2009; Mendoza-Poudereux et al., 2014). It is highly noticeable that metabolic engineering approaches to modulate expression of genes of the essential oil precursor pathway and those with relevance for oil composition have successfully increased the content of some target compounds (Dudareva and Pichersky 2008; Lange et al., 2011; Mendoza-Poudereux et al., 2014). Nevertheless, transgene expression levels overexpresing terpene synthases which lead to an increase in the content of certain terpenes seems not to entail changes in the essential oil yield (Lange et al., 2011; Mendoza-Poudereux et al., 2015). The yield production of essential oils appears more related to environmental and developmental factors and their interrelationships rather than under genetic control which it seems is linked more specifically to the chemical composition. Accordingly, several studies show that the variation in chemical composition of essential oils is mainly determined by genetic factors, emphasizing also the potential influence of other aspects such as environmental factors (Peixoto et al., 2015; Shojaiefar et al., 2015). Low genetic control for essential oil content and in turn chemical polymorphism strongly linked to concrete genotypes has been recently describe for *Lippia sidoides* where the content of main monoterpenes thymol and carvacrol was caused by genotypes (Dos Santos et al., 2015).

The Importance of the Plant Age

Many species with commercially important essential oils and others with recognized properties are perennial plants which means that they may live and be productive throughout several years. Studies have shown that the aging process of plants may cause a reduction of the essential oil content (Dos Santos et al., 2015) whereas other studies have shown that age can promote increased levels (Usano-Alemany et al., *in press*) and even in others with no significant effects observed (Cheng et al., 2009). Increasing or decreasing essential oil contents over time may lead to misevaluations of data regarding phenology of the plants.

Recent advances in this field focus mainly on optimizing crops and harvesting times with higher yield production and may provide high value information for optimal management.

Phenological Changes in the Chemical Composition of the Essential Oils

Implications of the Variability on the Product Final Quality

The full range of biological properties of the essential oils is linked to concrete chemical compositions which have shown efficacy in the experiments. These properties are usually attributed to the major components of the essential oils. However, in such a complex mixture, the possible effects or synergistic and antagonistic interactions between compounds are not negligible (Bakkali et al., 2008; Dhami and Mishra 2015). Phytochemical variation is a common phenomenon in aromatic and medicinal species. As a consequence of this fact, and unlike synthetic medicines, food preservatives or insecticides which are chemically consistent and stable, essential oils and other herbal products are rich and complex mixtures linked to inherent biological variation. In this regard it should be noted that in most cases, promising beneficial effects of essential oils are debased by spot samples or not well defined raw materials with lack of detail traceability regarding cultivation managements, collection times or edaphoclimatic conditions. To this extent, there is a definite necessity for chemical industries to manage high-quality botanicals, dealing with standardised or standardisable active compounds and matching the triple constraint 'quality–safety–efficacy' (Carrubba and Scalenghe 2012).

The first move towards this goal is to carry out a comprehensive and in-depth study of natural wild populations of potentially attractive market-based essential oil plants. These efforts have been notably intensive in the Mediterranean region where a vast number of species are well known by their essential oils. The content of the corresponding relevant secondary plant products in general is lower in plants grown in a moderate Atlantic climate than in those grown in semi-arid regions (Selmar and Kleinwächter, 2013b) and this is even more pronounced for the essential oils. The Mediterranean climate, with its typical warm dry summers and cool winters encompasses a variety of sub-climates mostly with drought periods which may last up to 5 or 6 months, during summer. Accordingly, natural epigenetic variations in natural systems allow the study of correlations between DNA methylation levels, environmental conditions and phenotypic expression (Lira-Medeiros et al., 2010). DNA methylation may thus affect phenotypic plasticity in response to adverse environmental conditions (Han and Wagner 2013), upregulating the production of diverse secondary metabolites. In this context, natural populations of most species are usually heterogenous and different essential oil profiles and volatile compositions can be discerned (Muñoz-Bertomeu et al., 2007; Ormeño et al., 2007; Rzepa et al.,2009; Herráiz-Peñalver et al., 2010, 2013; Rajabi et al., 2014; Usano-Alemany et al., 2014a). In the last decades, the description of variations in the chemical composition of essential oils in wild populations has been the subject of intensive research. The major monoterpenes, sesquiterpenes and even diterpenes show pronounced dynamics during vegetative cycle that have been confirmed in numerous species and under different geographical conditions (Table 4). As an example, the phenological investigation of wild populations of *Hypericum triquetrifolium* showed that the main essential oil compounds varied with the phenological stage (Hosni et al., 2011). The oils collected at the vegetative stage were found to be rich in both aliphatic and branched hydrocarbons with oxygenated compounds with mono and sesquiterpenes less represented. At the flowering stage, the oils consisted mainly of sesquiterpene hydrocarbons whilst at the fruiting stage the essential oils were distinguished by their higher content of oxygenated compounds (mono and sesquiterpenes). Sesquiterpenes were also higher in wild collected flowering plants of *Thymus algeriensis* (Zouari et al., 2012). Similar variations were found in *Hypericum perforatum* when wild populations were subjected to different collection times (Schwob et al., 2004). Higher content of oxygenated compounds during fruit maturation-vegetative has also been reported for *Salvia lavandulifolia* (Usano-Alemany et al., 2014b). Nonetheless, in *Thymus caramanicus* oxygenated compounds outstanding as the major portion throughout the whole phenological cycle (Nejad Ebrahimi et al., 2008).

Table 4. Essential oil main compounds reported from different phenological stages in representative wild and cultivated aromatic and medicinal plants

Species (a)	Main essential oil components	Phenology*				Strength of differences	ISO Norm
		Vegetative	Floral budding	Flowering	Fruiting		
1.	β-Caryophyllene	7.3	16.8	18.3	9.1	Weak	No
	Caryophyllene oxide	15.3	17.5	15.9	18.5		
2.	*n*-Octane/α-Pinene	15.2/10.7	-	17.2/13.2	13.2/9.7	Strong	No
	β-Caryophyllene	8.6		14.2	7.6		
3.1.	α-Pinene/1,8-Cineole/Camphor	-	-	13.0 /20.4/ 17.8	14.3/19.5/18.0	Weak	1342:2012
3.2.	α-Pinene/1,8-Cineole/Camphor	12.4/23.6/10.3	-	5.4/17.7/10.7	5.4/21.2/8.0	Weak	
3.3.	α-Pinene/1,8-Cineole/Camphor	11.1/17.6/24.6	-	-	13.5/6.4/27.9	Weak	
4.	Carvacrol/γ-Terpinene	58.9/8.0	66.9/6.4	68.9/4.6	60.2/6.7	Weak	No
5.	α-Pinene/1,8-Cineole/Camphor	10.2/15.0/14.3	-	8.5/13.7/12.0	-	Strong	No
6.	*p*-Cymene /γ-Terpinene/Thymol	33.0/12.6/16.1	-	30.4/14.3/26.2	40.8/9.7/23.3	Strong	No
7.1.	1,8-Cineole/Terpenyl Acetate	36.4/18.7	-	30.8/23.4	30.0/24.4	Strong	No
7.2.	Thymol/Carvacrol	-	48.8/0.5	40.7/0.9	41.0/1.1	Strong	
8.	Carvacrol/γ-Terpinene	72.6/6.0	-	76.6/5.0	45.4/6.0	Weak	No
	p-Cymene	4.7		5.0	6.3		
9.	L-Linanool/β-	5.6/21.9	4.1/22.4	14.9/23.0	7.3/30.6	Weak	4731:2012
	Citronellol/Geraniol	11.0	14.0	15.6	15.2		
10.1	α-Pinene/β-Pinene-Myrcene	6.6/13.2	7.2/16.0	10.0/18.1	6.1/13.0	Strong	3526:2015
	1,8-Cineol/Camphor	39.8/12.0	22.7/6.5	31.8/6.5	37.2/10.3		
10.2	α-Pinene/β-Pinene	10.9/9.7	-	7.5/11.8	-	Weak	
	1,8-Cineole/Camphor	25.2/11.0		31.3/15.6			
10.3	α-Pinene/β-Pinene-Myrcene	5.7/15.3	-	5.8/17.0	4.3/18.2	Strong	
	1,8-Cineole/Camphor	33.1/3.4		31.0/4.5	47.0/1.3		
11.1	α-Thujone/β-Thujone /Camphor	40.1/4.1/4.1	48.3/5.8/8.3	46.5/5.6/8.0	39.5/4.7/12.0	Weak	9909:2015
	Viridiflorol	12.3	12.6	4.3	13.4		
11.2	1,8-Cineole/α-Thujone /Borneol	16.8/12.8/8.4	19.3/13.2/6.9	22.3/9.1/11.0	19.5/9.1/14.6	Strong	
	β-Caryophyllene	10.5	10.6	4.9	7.3		
11.3	α-Thujone/β-Thujone /Camphor	39.5/4.8/6.7	-	33.1/4.1/6.0	41.3/5.6/6.3	Strong	
12.	Methyl chavicol or Estragole	81.5	87.5	89.3	88.5	Weak	No
13.	1,8-Cineole/Linalool /Camphor	-	-	32.6/30.4/13.3	29.1/38.2/15.3	Weak	3526:2015
14.	1,8-Cineole/Borneol /Camphor	31.8/20.0/13.8	-	13.6/10.3/4.3	17.3/14.0/15.3	Strong	3515:2012

Table 4. (Continued)

Species (a)	Main essential oil components	Phenology*				Strength of differences	ISO Norm
		Vegetative	Floral budding	Flowering	Fruiting		
15.	Geraniol/Geranyl acetate	-	57.5/28.6	68.0/14.2	75.0/5.56	Strong	4727:2014
16.1	Sabinene/*E*-β-Ocimene *allo*-Ocimene	26 /17 32†	153 /75 129†	599/119 161†	267/242 192†	Strong	13171
16.2	Carvacrol/*p*-Cymene	64.5/7.6	-	83.3/3.2	-	Strong	
17.	Terpinen-4-ol/Geranyl acetate *cis*-Sabinene hydrate	31.8/6.1 26.1	32.5/5.9 19.9	29.1/37.0 24.6	-	Weak	No
18.	Neral/Geranial	26.1/33.3	26.0/32.7	25.1/32.9	-	Weak	No
19.	β-Pinene/Pinocamphone	20.0/39.2	25.7/29.6	8.5/54.0	7.1/57.5	Strong	9841:2013
20.1 **20.2**	*p*-Cymene/Thymol/Carvacrol *p*- Cymene/Thymol/Carvacrol	3.9/15.5/36.0 3.8/11.0/52.4	- -	15.2/5.1/30.9/ 15.2/5.4/26.2	24.5/2.5/22.6 25.6/2.6/16.1	Strong Strong	No
21.	α-Thujone/β-Thujone	55.4/20.3	73.7/15.6	79.3/13.7	-	Strong	No

(a) Species correspond to those reported in Table 2.

*All data presented are referred to mean values published as %, †mg kg^{-1}.

Table 5. Phenological variation in the content (%) of 1,8-cineole in the essential oil of cultivated clonal lines of *Salvia lavandulifolia*

COMPOUND			Harvesting events in one year					
			I	**II**	**III**	**IV**	**V**	**VI**
	Clones	***HSD***	***A***	***C***	***B***	***B***	***AB***	***AB***
1,8-cineole	A2	*de*	37.9	**24.5**	**13.3**	**24.5**	**25.2**	**18.3**
	A3	*e*	**25.5**	**17.7**	**18.4**	**11.6**	**16.1**	**5.7**
	B1	*bc*	53.1	**16.8**	38.2	42.7	46.7	43.7
	B2	*bc*	47.9	**17.5**	41.0	43.2	44.3	42.7
	B3	*ab*	51.8	**27.8**	43.3	49.4	48.8	57.0
	C1	*de*	**29.6**	**13.8**	**17.3**	**24.9**	38.6	**31.6**
	C2	*e*	**24.2**	**18.2**	**11.3**	**8.2**	**15.6**	**15.0**
	C3	*ab*	54.1	**31.3**	40.6	42.1	56.2	52.1
	D1	*a*	62.7	48.8	52.5	49.1	56.8	54.3
	D2	*cd*	46.0	**22.0**	**24.3**	34.4	37.2	**31.2**
	D3	*abc*	49.5	**10.8**	49.2	50.6	47.5	43.0
Average per harvesting event			*43.8*	*22.7*	*31.8*	*34.6*	*39.4*	*35.9*

Cells in bold indicate samples from group 1 with low content of 1,8-cineole (μ = 19.74%). The rest of cells belong to group 2 with high content of 1,8-cineol (μ = 47.13%). The methodologies used were the k-mean and Mahalanobis distance for the group separation and for the setting of groups differences, respectively, by using the major 15 compounds of the essential oil in cultivated plants. Different uppers and lower cases in the row and in the column, respectively, indicate significant differences in relative percentage ($p < 0.05$). The phenological stages were the following: I (vegetative), II (new leaves formation), III (flowering), IV (late flowering) V (fruit maturation), VI (vegetative II). Table extracted from Usano-Alemany et al. (2014b) with modifications.

The evolution of main terpenoids according to the phenology of the plants is highly complex and difficult to clarify due to the proved heterogeneity. Chemical profiles may be different between species and also within species regarding different locations and growing conditions (Tounsi et al., 2011). As previously mentioned, several factors can influence the qualitative and quantitative changes in the chemical profile and they must be handled accordingly in each particular case. In this regard, cultivation of medicinal and aromatic plants is presented as the best way to reduce these sources of variation generating material grown under common conditions in contrast to wild collection. This fact would help to solve the lack of reproducibility of plant-based products and to achieve assurance in quality and traceability. However, to minimize the above mentioned uncertainties, an appropriate characterization of propagating material is needed in order to promote a better cultivation management.

Genetic variation between plants can be a major source of variation (Gorelick and Nirit 2014) in plant secondary metabolites does so for essential oils, and it can considerably affect the amounts and type of metabolites produced. Accordingly, studies carried out with particular genotypes have shown great stability on their chemical profiles hence the investigation on cultivated clones has been proved to be the best method for monitoring real phenological changes in the composition of the essential oils (Jordán et al., 2006; Shojaiefar et al., 2015; Smitha and Virendra 2015; Usano-Alemany., *in press*). Monitoring provides valuable information on optimum harvesting period and real phenological influences on essential oil chemical composition which is a key reporting for commercial growers (Table 5). Accordingly, 1,8-cineole showed its lower content in a clear manner in *Salvia lavandulifolia* when new leaves begin to grow in Spring even though the relative percentage is remarkable linked to concrete clonal lines (Usano-Alemany et al., 2014b). Some clonal lines had low values throughout the phenology. Therefore, chemical variability and lack of reproducibility are the primary reasons why essential oils and other plant extracts are not yet utilized on a large scale even though a plethora of new bioactivities and uses are described each year (Christaki et al., 2012; Hyldgaard, 2012; Amorati et al., 2013; Ellse and Wall 2014).

Genetic Control of the Essential Oil Composition

Levels of different terpenes of the essential oils are thought to be controlled transcriptionally at the levels of monoterpene and sesquiterpene synthases although to a different extent with the amount of some compounds which are controlled by post-transcriptional regulation (Degenhardt et al., 2009; Grausgruber-Gröger et al., 2012). Different content in monoterpenes, sesquiterpenes and diterpenes during phenology might be due to the overexpression of a particular terpene synthase gene, which in some cases showed highly remarkable differences between plant organs (Muñoz-Bertomeu et al., 2008; Mendoza-Poudereux et al., 2014). For example, linalool accumulation in *Lavandula angustifolia* was correlated with transcript levels for endogenous linalool synthase (LIS) gene (Lane et al., 2010). Transgenic plants for this gene maintained the high content at all developmental stages as compared with control plants. This high linalool producing phenotype was maintained in leaf essential oil of the progenies that inherited the transgene. This phenotype was, however, less evident in the essential oil from flowers of these progenies. This fact suggests a strong spatial regulation of monoterpene shynthases (Mendoza-Poudereux et al., 2014; 2015).

Different Organs and Essential Oil Composition

Results available regarding the influence of phenology on essential oil chemical composition are often dissenting and even with conflicting results and are highly dependent on the species (Table 4). The phenological variation has been reported as highly noticeable both in wild collected plants and in cultivated plant material. Nonetheless, cultivation provides more clarity and easier enforcement to understand the behaviors and patterns. Some aspects must be taken into account when referring to phenological changes in the essential oil components isolated from distillation of aerial parts. First of all, different developmental stage involve different plant parts collected and with different harvesting ratios. Flowers, new leaves, growing steams and fruits can be collected depending on the phenology. The irregular allocation of essential oil compounds has been suggested in several studies (Bertoli et al., 2004; Bettaieb et al., 2010; Tounsi et al., 2011). Flowers of numerous species have been reported with remarkable different composition compared with leaves (Table 6). The level and activity of various terpene synthases involved in the production of essential oils is presented differently between plant parts and also between species considering the same organ. The methyl-erythritol phosphate (MEP) pathway which gives rise to the formation of monoterpenes and diterpenes and the cytosolic mevalonate (MVA) pathway from which sesquiterpenes result may be boosted differentially between different plant organs (Degenhardt et al., 2009). The content of sesquiterpenes was reported in a remarkable higher content in flowers of *Rosmarinus officinalis* and *Psoralea bituminosa* with a large presence of β-caryophyllene, ß- farnesene, and α-humulene, reaching up to 45% and 80% of total identified compounds, respectively (Bertoli et al., 2004; Yosr et al., 2013). Similarly, fragrance from *Cananga odorata* due to its flowers is mainly produced by the predominance of sesquiterpenes in its chromatographic profile mainly by the involvement of sesquiterpene synthases (Jin et al., 2015). On the other hand, flowers of *Eucalyptus oleosa* and *Salvia officinalis* showed high percentage of 1,8-cineole as major oxygenated monoterpene (Perry et al., 1999; Ben Marzoug et al., 2011). In flowers of *Pelargonium graveolens* oxygenated monotepenes, represented 70.3% of the total essential oils, β-citronellol, geraniol and δ-selinene being the dominant components (Boukhris et al., 2013). The chemical composition of the essential oil from flowers of *Salvia lavandulifolia* (Spanish sage) was remarkably different to that described for leaves and stems and in turn totally different compared with its close species *S. officinalis*. This essential oil mostly contains linalool (36.9%) and linalyl acetate (17.9%) which were only described as traces in leaves and stems being in turn characterized by 1,8-cineole, camphor, α-pinene, β-pinene and limonene throughout the whole phenology (Usano-Alemany 2012). These two main compounds of flowers are also the most representative ones in essential oils of other species as lavender (*Lavandula angustifolia*) and spike lavender (*Lavandula latifolia*) (only linalool), species that share habitat with Spanish sage and whose essential oil are basically obtained from their inflorescences. Hence a distinct ecological and functional character of the volatiles is suggested in that case. Another interesting case of a very polarized chemical composition between plant parts is given by *Eryngium* species. Some species like *E. corniculatum*, *E. campestre* or *E. planum* provide high content of valuable compounds such as phyllocladenes and falcarinol in their essential oils isolated only from roots (Thiem et al., 2001; Palá-Paúl et al., 2008, 2010).

In the above mentioned case of *Lavender* species and many others, in which the isolation of essential oils is undertaken by distillation-extraction of concrete parts, the chemical composition is given by the content in this organ. Nonetheless, species in which whole aerial parts are subjected to distillation, the relation between leaves, stems and flowers is an important fact to consider when phenology is evaluated. Volatile organic compounds are important components of plant´s chemical phenotype and their ecological relevance has been raised for decades (Maffei et al., 2011). Leaves, flowers, fruits and roots produce them as pollinator-atractors and seed dispersers, defense against herbivores and pathogens and for signaling involvement in plant-plant communication. These ecological roles determine their different distribution within plant organs and eventually, may have major implications when aerial parts are subjected to distillation. It is therefore very important for the evaluation and reporting of phenological studies to clarify this relationship between plant organs when aerial parts are collected for distillation.

Table 6. Essential oil main compounds reported from different organs in representative wild and cultivated aromatic and medicinal plants

Species (a)	Main essential oil components	Plant part*				Strength of differences
		Leaves	Steams	Flowers	Fruits	
1.	2-Nonanol/Trans-Linanool oxide 2-Undecanone	36.5/12.6 35.0	34.2/1.3 30.0	28.4/5.0 52.3	6.7/18.4 20.6	Strong
2.	p-Cymene/γ-Terpinene/Carvacrol	6.3/5.4/54.1	-	3.5/6.3/77.6	5.4/5.4/76.7	Strong
3.	α-Pinene/1,8-Cineole/Camphor Borneol	10.1/35.8/14.5 9.2	0.7/5.8/4.9 9.6	1.3/10.3/4.4 11.9	-	Strong
4.	α-Pinene/β-Caryophyllene Germacrene D	0.3/23.2 24.3	-	2.5/18.1 18.2	50.2/6.5 7.0	Strong
5.	β-Pinene/Bornyl Acetate/γ-Terpinene	0.51/5.27/1.62	13.6 /23.8 /-	2.0/0.2/51.7	-	Strong
6.	β-Pinene/Pinocarvone *cis*-Pinocamphone	9.8/7.7 57.7	-	7.8/24.9 49.7	5.7/5.5 57.3	Weak
7.	Germancrene D/Bicyclogermacrene Geranyl formate/γ-Muurolene	39.2/33.0 -/-	-	-	-/- 27.3/10.3	Strong
8.	1,8-Cineole/*trans*-Pinocarveol Spathulenol/γ-Eudesmol	8.7/- 16.1/15.0	31.5/9.9 3.5/5.6	47.0/0.1 -/12.5	29.1/0.1 3.4/16.4	Strong
9.	Thujones/β-pinene	31.0/9.9	37/7.6	16.0/27.1	-	Weak
10.	Limonene/*cis*-Chrysanthenyl acetate Bornyl acetate/(*Z*)-Falcarinol [3.2/- / -/64.4] (*in roots*)	14.7/- 4.6/-	11.3/- 18.1/0.2	0.9/43.2 -/0.4	-	Strong

(a) Species correspond to those reported in Table 3. *All data presented are referred to mean values published as %.

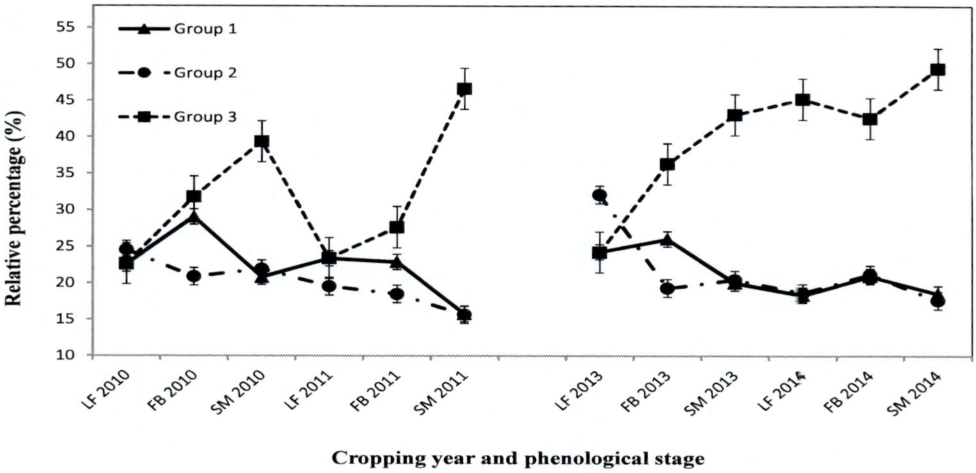

Figure 2. Temporal patterns of the essential oil chemical composition of *Salvia lavandulifolia* throughout four years of cultivation. Three groups of correlated compounds are presented. Group 1 comprises α-pinene, the sum of β-pinene + myrcene, group 2 includes camphene, limonene, camphor, (*E*)-β-caryophyllene and caryophyllene oxide and group 3 corresponding to 1,8-cineole. LF: leaves formation, FB: Full bloom, SM: seeds maturation.

Another important factor is the distribution and the total number of glandular trichomes. Once the essential oil is formed and accumulated, it is stored in the oil glands for a long time. An analysis comprising a mixture of leaves of different age would not allow us to estimate the influence of environmental conditions (Grausgruber-Gröger et al., 2012) and might be behind some of the chemical differences found according to the phenology. Essential oils composition is highly related to the glandular trichomes distribution, morphology and anatomy. That is the case of *Mentha* × *piperita* in which there was a strong correlation with differences in monoterpenoids (Rios-Estepa et al., 2010). The outcome of monoterpene and sesquiterpene synthases activity is stored in these epidermial glands and hence their location and developmental stage is crucial for the subsequent quality that can be obtained.

Plant Aging vs Composition

Many species with commercially important essential oils and others with recognized properties are perennial plants, which means that they may live and be productive throughout several years. Changes in essential oil chemical composition according to the plant aging may have remarkable consequence upon the phenological characterization. The age and developmental stage of the plant can redirect the metabolic pathway and lead to the biosynthesis of different compounds that might cause an increase of certain components. The relatively high proportion of (E)-β-caryophyllene in young plants suggests that this may be one compound that is differentially induced during the earlier stages of plant ontogenetic development in some species (Hare 2010). An aspect that appears to be rather common among long-term views and monitoring of essential oil production is that the content of major compounds is increasingly relevant regarding the plant aging. This effect has been reported for some main essential oil components such as geraniol in palmorosa (*Cymbopogon martini*)

(Smitha and Virendra 2015) and 1,8-cineole in *Mentha pulegium* (Rodrigues et al., 2013). The latter compound is shown to play also an increasingly important role over time in the essential oil profiles of different genotypes of *Salvia lavandulifolia* (Figure 2). Nonetheless, in other cases the chemical polymorphism with plant aging in relation to the content of main compounds such as thymol and carvacrol is highly related to the genotypes (Dos Santos et al., 2015).

CONCLUSION

Plant phenological changes influence the production and chemical composition of essential oils. These changes affect to the ontogeny and developmental stage of the different plant organs and to the distribution, density and total number of glandular trichomes, aspects that directly affect to the biosynthesis of essential oils. At the same time, phenological development is related to other external factors such as seasonal and environmental changes and possible stressful situations which have been reported to clearly influence essential oil yield and may mask the effects of phenology itself. On the other hand, the impact of phenology on chemical composition has been also described, although in this concern the genetic inheritance of the plant seems to be determining in relation to the chemical profile of the essential oils. To conclude, a better knowledge about the set of factors affecting chemical composition and yield of essential oil bearing plants is crucial in order to ensure homogeneous and standardizable productions of medicinal and aromatic species with a recognized quality and linked to the biological activity expected from them.

REFERENCES

Abreu, Nacif. Ilka., Porto, L. M., Andre, Marsaioli., Anita, J. and Mazzafera, Paulo. (2004). "Distribution of bioactive substances from *Hypericum brasiliensis* during plant growth". *Plant Science*, *167*, 949-954. doi: 10. 1016/ j.plantsci.2004.06.002.

Adams, Robert. P. (2010). "Chemosystematics of *Juniperus*: Effects of leaf drying on essential oil composition". *Phytologia*, *92*, 186-198.

Amorati, Riccardo., Foti, Mario. C. and Valgimigli, Luca. (2013). "Antioxidant activity of the essential oils". *Journal of Agricultural and Food Chemistry*, *61*, 10835-10847. doi:10.1021/jf403496k.

Andrew, Rose. L., Keszei, Andras. and William, Foley. J. (2013). "Intensive sampling identifies previously unknown chemotypes, population divergence and biosynthetic connections among terpenoids in *Eucalyptus tricarpa*". *Phytochemistry*, *94*, 148-158. doi: 10.1016/ j.phytochem. 2013. 05.002.

Arraiza, Maria. P., Arrabal, Carlos. and José, López. V. (2012). "Seasonal Variation of Essential Oil Yield and Composition of Sage (*Salvia offcinalis* L.) Grown in Castilla-La Mancha (Central Spain)". *Notulae Botanicae Horti Agrobotanici Cluj-Napoca*, *40*, 106-108. doi:10.15835/ nbha4028311.

Badawy, E. I. Mohammed, El-Arami., Sailan, A. A. and Samir Abdelgaleil, A. M. (2010). "Acaricidal and quantitative structure activity relationship of monoterpenes against the two-spotted spider mite *Tetranychus urticae*". *Experimental and Applied Acarology. 52*, 261-274. doi: 10.1007/s10493-010-9363-y.

Bakkali, F., Averbeck, S., Averbeck, D. and Idaomar, M. (2008). "Biological effects of essential oils – A review". *Food and Chemical Toxicology*, *46*, 446-475. doi:10.1016/j.fct.2007.09.106.

Barakauskienė, Renata., Venskutonis, Petras. Rimantras., Dambrauskienė, Edita. and Pranas Viškelis, P. (2013). "Harvesting time influences the yield and oil composition of *Origanum vulgare* L. ssp. *vulgare* and ssp. *Hirtum*". *Industrial Crops and Products*, *49*, 43-51. doi:10.1016/ j.indcrop.2013.04.024

Baran, Pelin., Özdemir, Canan. and Kamuran, Aktas. (2010). "Structural investigation of glandular trichomes of *Salvia argentea*". *Biologia*, *65*, 33-38. doi: 10.2478/s11756-009-0209-1.

Barra, Andrea. (2009). "Factors affecting chemical variability of essential oils: a review of recent developments". *Natural Product Communications*, *4*, 1147-1154.

Béjaoui, Afef., Chaabane, Hédia., Maroura, Jemli., Boulila, Abdennacer. and Mohamed, Boussaid. (2013). "Essential oil composition and antibacterial activity of *Origanum vulgare* subsp. *glandulosum* Desf. at different phenological stages". *Journal of Medicinal Food*, *16*, 1115-20. doi: 10.1089/jmf.2013.0079.

Ben, Jemâa., Jouda, Mediouni., Haouel, Soumaya., Bouaziz, Mohamed. and Mohamed Larbi, Khouja. (2012). "Seasonal variations in chemical composition and fumigant activity of five *Eucalyptus* essential oils against three moth pest of stored dates in Tunisia". *Journal of Stores Products Research*, *48*, 61-67. doi: 10.1016/j.jspr.2011.10.001.

Ben Marzoug, Nacer. H., Romdhane, Mehrez., Lebrihi, Ahmed., Mathieu, Florence., Couderc, François., Abderraba, Manef., Khouja, Larbi. M. and Jalloul, Bouajilla. (2011). "*Eucalyptus oleosa* essential oils: chemical composition and antimicrobial and antioxidant activities of the oils from different plant parts (stems, leaves, flowers and fruits)". *Molecules*, *16*, 1695-1709. doi: 10.3390/molecules16021695.

Bertoli, Alessandra., Menichini, Francesco., Noccioli, Cecilia., Morelli, Ivano. and Luisa, Pistelli. (2004). "Volatile constituents of different organs of Psoralea bituminosa L." *Flavour and Fragance Journal*, *19*, 166-177. doi:10.1002/ffj.1315.

Bettaieb, I., Nakhama, N., Aidi Wannes, W., Kchouk, M. E. and Marzouk, B. (2009). "Water deficit affects on *Salvia officinalis* fatty acids and essential oils composition". *Scientia Horticulturae*, *210*, 271-275. doi:10.1016/ j.scienta.2008. 10.016.

Bettaieb, Innes., Bourgou, Soumaya., Wannes, W., Aidi, Hamrouni., Ibtissem, Limam. Ferid. and Brahim, Marzouk. (2010). "Essential oils, phenolics and antioxidant activities of different parts of cumin (*Cominum cymimum* L.)". *Journal of Agricultural and Food Chemistry*., *58*, 10410- 10418. doi: 10. 1021/jf102248j.

Böszörményi, Andrea., Héthelyi, Éva., Farkas, Ágnes., Horváth, Györgyi., Papp, Nora., Lemberkovics, Eva. and Éva, Szoke. (2009). "Chemical and genetic relationships among sage (*Salvia officinalis* L.) cultivars and Judean sage (*Salvia judaica* Boiss.)". *Journal of Agricultural and Food Chemistry*, *57*, 4663-4667.

Boukhris, M., Bouaziz, M. and Sayadi, S. (2013). "Comparative essential oil composition and antioxidant activity of flowers, leaves and stems of rose-scented geranium (*Pelargonium Graveolens* L'Hér.) from Tunisia". *Acta Horticulturae*, *997*, 111–116. doi:10.17660/ActaHortic.2013.997.13

Calo, Juliany Rivera., Crandall, Phillip. G., O' Bryan, Corliss. A. and Ricke, Steven. C. (2015). "Essential oils as antimicrobial in food systems - a review". *Food Control, 54*, 111-119. doi:10.1016/j.foodcont.2014.12.040

Carrubba, Alessandra. and Riccardo, Scalenghe. (2012). "The scent of Mare Nostrum: medicinal and aromatic plants in Mediterranean soils". *Journal of the Science of the Food and Agriculture*, *92*, 1150-1170. doi: 10.1002/ jsfa.5630.

Cheng, Sen-Sung., Chua, Meng-Thong., Chang, Ed-Haun., Huang, Chin-Gi., Chen, Wei-June. and Shang-Tzen, Chang. (2009). "Variations in insecticidal activity and chemical compositions of leaf essential oils from *Cryptomeria japonica* at different ages". *Bioresource Technology*, *100*, 465-470. doi: 10.1016/j.biortech.2007.11.060.

Christaki, Efterpi., Bonos, Eleftherios., Giannenas, Ilias. and Panagiota. Florou-Paneri. (2012). "Aromatic plants as a source of bioactive compounds". *Agriculture*, *2*, 228-243. doi:10.3390/agriculture2030228.

Cordoba, Elisabeth. Salmi, Mari. and Patricia, León. (2009). "Unravelling the regulatory mechanisms that modulate the MEP pathway in higher plants". *Journal of Experimental Botany*, *60*, 2933–2943. doi: 10.1093/ jxb/erp190.

Da Silva, Erica Biolcati. P., Soares, Marisi. G., Mariane Bruna, Vallim, Marcelo. A., Pascon, Renata. C., Sartorelli, Patricia. and João Henrique, Lago. G. (2013). "The seasonal variation of the chemical composition of essential oils from *Porcelia macrocarpa* R.E. Fries (Annonaceae) and their antimicrobial activity". *Molecules*, *18*, 13574-13587. doi:10.3390/ molecules181113574.

Degenhardt, Jörg., Köllner, Tobias G. and Gershenzon, Jonathan. (2009). "Monoterpene and sesquiterpene synthases and the origin of terpene skeletal diversity in plants". *Phytochemistry*, *70*, 1621-1637. doi: 10.1016/j.phytochem.2009.07.030.

De Luna, Bruna Nunes., Antunes e Defaveri, Anna Carinna., Sato, Alice., Ribeiro Bizzo, Humberto., Freitas, Maria de Fátima. and Claudia, Franca Barros. (2014). "Leaf secretory tissues in Myrsine coriacea and Myrsine venosa (Primulacea): ontogeny, morphology and chemical composition of the essential oils". *Botany*, *92*, 757-766. doi: 10.1139/cjb-2014-0044.

De Vicenci, M., Silano, M., De Vicenci, A., Maialetti, F. and Schazzocchio, B. (2002). "Constituents of aromatic plants: Eucalyptol". *Fitoterapia*, *73*, 269-275. doi:10.1016/S0367-326X(02)00062-X.

Dhami, Namraj. and Akkal Dev, Mishra. (2015). "Phytochemical variation: How to resolve the quality controversies of herbal medicinal products?" *Journal of Herbal Medicine*, *5*, 118-127. doi:10.1016/ j.hermed.2015.04.002.

Dorman, H. J., Damien, Koşar., Kahlos, Müberra., Kirsti, Yvonne, Holm. and Raimo, Hiltunen. R. (2003). "Antioxidant properties and composition of aqueous extracts from *Mentha* species, hybrids, varieties, and cultivars". *Journal of Agricultural and Food Chemistry*, *51*, 4563-4569. doi 10.1021/ jf034108k.

Dos Santos, C. P., Oliveira, T. C., Oliveira Pinto, J. A., Fontes, S. S., Oliveira Cruz, E. M., Arrigoni-Blank, M. F., Andrade, T. M., Matos, I. L., Alves, P. B., Innecco, R. and Arie Fitzgerald Blank. (2015). "Chemical diversity and influence of plant age on the essential oil *from Lippia sidoides* Cham. Germplasm". *Industrial Crops and Products*, *76*, 416-421. doi:10.1016/ j.indcrop.2015.07.017.

Dudareva, Natalia. and Eran, Pichersky. (2008). "Metabolic ingineering of plant volatiles". *Current opinion in Biotechnology*, *19*, 181-189. doi 10.1016/ j.copbio.2008.02.011.

Nejad Ebrahimi, S., Hadian. J., Mirjalili, M. H., Sonboli, A. and Yousefzadim, M. (2008). "Essential oil composition and antibacterial activity of *Thymus caramanicus* at different phenological stages". *Food Chemistry*, *110*, 927-931. doi:10.1016/ j.foodchem.2008.02.083.

Elaissi, Ameur., Hadj, Salah., Karima, Hadj., Mabrouk, Samia., Larbi, Khouja., Mohamed, Chemli. Rachid. and Fethia, Harzallah-Skhiri. (2011). "Antibacterial activity and chemical composition of 20 *Eucalyptus* species' essential oils". *Food Chemistry*, *129*. 1427-1434. doi:10.1016/ j.foodchem.2011.05.100

Eldridge, K. G., Davidson, J., Harwood, C. E. and van Wyk, G. (1993). *Eucalypt domestication and breeding*. Oxford University Press. Oxford.

Ellse, L. and Wall, R. (2014). "The use of essential oils in veterinary ectoparasite control: a review". *Medical and Veterinary Entomology*, *28*, 233-243. doi: 10.1111/mve.12033.

Figueiredo, A., Cristina, Barroso., José, G., Pedro, Luis G. and Johannes, J. C. Scheffer. (2008). "Factors affecting secondary metabolite production in plants: volatile components and essential oils". *Flavour and Fragance Journal*, *23*, 213-226. doi: 10.1002/ffj.1875.

Gechev, Tsanko S., Hille, Jacques., Woerdenbag, Herman J., Benina, Maria., Mehterov, Nikolai., Toneva, Valentina., Fernie, Alisdair. R. and Bernd, Mueller-Roeber. (2014). "Natural products from resurrection plants: potential for medical application" *Biotechnology advances*, *32*, 1091-1101. doi: 10-1016/j.biotechadv.2014.03.005.

Gershenzon, Jonathan. and Natalia, Dudareva. (2007). "The function of terpene natural products in the natural world". *Natural Chemical Biology*, *3*, 408-14. doi:10.1038/nchembio.2007.5.

Glas, J., Joris, Schimmel., Bernardus, C. J., Schimmel, Alba., Juan, M., Escobar-Bravo, Rocío., Schuurink, Robert. C. and Merijn, Kant. R. (2012). "Plant Glandular trichomes as targets for breeding or engineering of resistance to herbivores". *International Journal of Molecular Sciences*, *13*, 17077-17103. doi:10.3390/ijms131217077.

Gnavi, Giorgio., Bertea, Cinzia. M., Usai, Marianna. and Massimo, Maffei. E. (2010). "Comparative characterization of *Santolina insularis* chemotypes by essential oil composition, 5S-rRNA-NTS sequencing and EcoRV RFLP-PCR". *Phytochemistry*, *71*, 930-936. doi: 10.1016/ j.phytochem. 2010.03.012.

Gonzalez-Coloma, Azucena., Lopez-Balboa, Carmen., Santana, Omar., Reina, Matías. and Braulio, Fraga. M. (2011). "Triterpene-based plant defences". *Phytochemistry Reviews*, *10*, 245-260. doi:10.1007/s11101-010-9187-8.

Goodger, Jason Q. D. and Woodrow, Ian. E. (2009). "The influence of ontogeny on essential oil traits when micropropagating *Eucalytus polybractea*". *Forest Ecology and Management*, *258*, 650-656. doi:10.1016/ j.foreco. 2009.04.033.

Gorelick, Jonathan. and Nirit, Bernstein. (2014). "Elicitation: an underutilized tool in the development of medicinal plants as a source of therapeutic secondary metabolites". *Advances in Agronomy*, *124*, 201-230. doi:10. 1016/B978-0-12-800138-7.00005-X.

Grausgruber-Gröger, Sabine., Schmiderer, Corinna., Steinborn, Ralf. and Johannes, Novak. (2012). "Seasonal influence of gene expression of monoterpene synthases in *Salvia officinalis* (Lamiaceae)". *Journal of Plant Physiology*, *169*, 353-359. doi: 10.1016/j.jplph.2011.11.004.

Grulova, Daniela., De Martino, Laura., Mancini, Emilia., Salamon, Ivan. and Vicenzo, De Feo. (2015). "Seasonal variability of the main components in essential oil of

Mentha × piperita L". *Journal of the Science of Food and Agriculture*, *95*, 621-927. doi: 10.1002/jsfa.6802.

Hamrouni Sellami, I., Maamouri, Emma., Chahed, Thouraya., Wannes, Wissem Aidi., Kchouk, Mohamed. Elyes. and Brahim, Marzouk. (2009). "Effect of growth stage on the content and composition of the essential oil and phenolic fraction of sweet marjoram (*Origanum majorana* L.)". *Industrial Crops and Products*, *30*, 395, 402. doi:10.1016/ j.indcrop.2009.07.010.

Han, Soon-Ki. and Wagner, Doris. (2013). "Role of chromatin in water stress responses in plants". *Journal of Experimental Botany*, *65*, 2785-2799. doi: 10.1093/jxb/ert403.

Harada, Emiko., Kim, Ji. Ah., Meyer, Andreas. J., Hell, Rüdiger., Clemens, Sthepan. and Yong-Eui, Choi. (2010). "Expression profiling of tobacco leaf trichomes identifies genes for biotic and abiotic stresses". *Plant and Cell Physiology*, *51*, 1627-1637. doi: 10.1093/pcp/pcq118.

Hare, J. Daniel. (2010). "Ontogeny and seasonal constrain in the production of herbivore-inducible plant volatiles in the field". *Journal of Chemical Ecology*, *36*, 1363-1374. doi: 10.1007/s10886-010-9878-z.

Herráiz-Peñalver, David., Usano-Alemany, Jaime., Cuadrado, Joaquín., Jordán, Maria José., Lax, J., Vanessa, Sotomayor., Jose, Antonio. and Jesús, Palá-Paúl. (2010). "Essential oil composition of wild populations of *Salvia lavandulifolia* Vahl. from Castilla-La Mancha (Spain)". *Biochemical Systematics and Ecology*, *38*, 1224-1230. doi:10.1016/ j.bse.2010.10.015.

Herráiz-Peñalver, David., Cases, Maria. Angeles., Varela, Federico., Navarrete, Paloma., Sánchez-Vioque, Raúl. and Jaime, Usano-Alemany. (2013). "Chemical characterization of *Lavandula latifolia* Medik. essential oil from Spanish wild populations". *Biochemical Systematics and Ecology*, *46*, 59-68. doi:10.1016/j.bse.2012.09.018.

Herráiz-Peñalver, David., Ortiz de Elguea-Culebras, G., Sánchez-Vioque, Raúl. and Omar, Santana-Méridas. (2015). "Identification of a hybrid species of sage (*Salvia officinalis L.* x *S. lavandulifolia* subsp. *lavandulifolia*) through the study of the essential oil". *Journal of Essential Oil Research*, *27*, 363-372. doi: 10.1080/10412905.2015.1031918.

Homer, Laura. E., Leach, David. N., Lea, David., Lee, Slade., Lee, Henry., Robert, J. and Peter, Baverstock. R. (2000). "Natural variation in the essential oil content of *Melaleuca alternifolia* Chell (Myrtaceae)". *Biochemical Systematics and Ecology*, *28*, 367-382. doi:10.1016/S0305-1978(99)00071-X.

Hosni, Karim., Msaada, Kamel., Ben Taârit, Mouna. and Brahim, Marzouk. (2011). "Phenological variations of secondary metabolites from *Hypericum triquetrifolium* Turra". *Biochemical Systematics and Ecology*, *39*, 43-50. doi:10.1016/j.bse.2011.01.001.

Howes, Melanie-Jayne. R. and Houghton, Peter. J. (2012). "Etnobotanical treatment strategies against Alzheimer disease". *Current Alzheimer research*, *9*, 67-85. doi: 10.2174/156720512799015046.

Hyldgaard, Morten., Mygind, Tina. and Rikke Louise, Meyer. (2012). "Essential oils in food preservation: mode of action, synergies, and interactions with food matrix components". *Frontiers in Microbiology*, *3*, 1-24. doi: 10.3389/fmicb.2012.00012.

Isman, Murray B., Miresmailli, Saber. and Cristina, Machial. (2011). "Commercial opportunities for pesticide based on plan essential oils in agriculture, industry and consumer products". *Phytochemistry Reviews*, *10*, 197-204. doi: 10.1007/s11101-010-9170-4.

Jamali, Chaim Alaoui., Kasrati, Ayaoub., Bekkouche, Khalid., Hassani, Lahce., Wohlmuth, Hans., Leach, David. and Abdelaziz, Abbad. (2013). "Phenological changes to the chemical composition and biologica activity of the essential oil from Moroccan endemic thyme (*Thymus maroccanus* Ball)". *Industrial Crops and Products*, *49*, 366-372. http:// dx.doi.org/ 10. 1016/j.indcrop.2013.05.016.

Janicsák, Gábor., Veres Katalin, Kakasy., András, Zoltán. and Imre, Mathe. (2006). "Study of the oleanolic and ursolic acid contents of some species of the *Lamiaceae*". *Biochemical Systematics and Ecology*, *34*, 392-396. doi: 10.1016/j.bse.2005.12.004.

Jin, Jingjing., Kim, Mi Jung., Dhandapani, Savitha., Tjhang, Jessica. Gambino., Yin, Jun-Lin., Wong, Limsoon., Sarojam, Rajani., Chua, Nam-Hai. and In-Cheol, Jang. (2015). "The floral transcriptone of ylang ylang (*Cananga odorata* var. fruticosa). uncovers biosynthetic pathways for volatile organic compounds and a multifunctional and novel sesquiterpen synthase". *Journal of Experimental Botany*, *66*, 3959-3975. doi: 10.1093/jxb/erv196.

Jordán, Maria José., Martínez, R. M., Goodner, K. L., Baldwin, E. A. and Jose, Antonio Sotomayor. (2006). "Seasonal variation of *Thymus hyemalis* Lange and Spanish *Thymus vulgaris* essential oils composition". *Industrial Crops and Products*, *24*, 253-263. doi: 10.1016/ j.indcrop. 2006.06.011.

Jordán, Maria José., Lax, Vanesa., Rota, Maria. C., Lorán, Susana. and Jose Antonio, Sotomayor. (2012). "Relevance of Carnosic Acid, Carnosol, and Rosmarinic Acid Concentrations in the in Vitro antioxidant and antimicrobial activities of *Rosmarinus officinalis* (L.) methanolic extracts". *Journal of Agricultural and Food Chemistry*, *60*, 9603-9608. dx.doi.org/10.1021/jf302881t.

Jordán, Maria José., Lax, Vanesa., Rota, Maria. C., Lorán, Susana. and Jose Antonio, Sotomayor. (2013). "Effect of the phenological stage on the chemical composition, and antimicrobial and antioxidant properties of *Rosmarinus officinalis* L. essential oil and its polyphenolic extract". *Industrial Crops and Products*, *48*, 144-152. doi: 10.1016/ j.indcrop. 2013.04.031.

Jordán, Maria José., Lax, Vanesa., Rota, Maria. C., Lorán, Susana. and Jose Antonio, Sotomayor. (2014). "Effect of thephenological stage on the chemical composition, and antimicrobial and antioxidant properties of *Rosmarinus officinalis* L essential oil and its polyphenolic extract". *Industrial Crops and Products*, *48*, 144-152. http:// dx.doi.org/ 10.1016/ j.indcrop.2013.04.031.

Karray-Bouraoui, Najoua., Rabhi, Mokded., Neffati, Manel., Baldan, Barbara., Ranieri, Annamaria., Marzouk, Brahim., Lachaâl, Mokhtar. and Abderraza, Smaoui. (2009). "Salt effect on yield and composition of shoot essential oil and trichome morphology and density on leaves of *Mentha pulegium*". *Industrial Crops and Products*, *30*, 338-343. doi: 10.1016/ j.indcrop. 2009.06.003.

Kennedy, David O. and Wightman, Emma. L. (2011). "Herbal extracts and phytochemicals: Plant secondary metabolites and the enhancement of human brain function". *Advanced in nutrition*, *2*, 32-50. doi: 10.3945/ an.110.000117.

Khadivi-Khub, Abdollah., Aghaei, Yaqoub. and Mohammad-Hossein, Mirjalili. (2014). "Phenotypic and phytochemical diversity among different populations of *Stachys lavandulifolia*". *Biochemical Systematics and Ecology*, *54*, 272-278. doi: 10.1016/j.bse.2014.02.024.

Khan, Reehana., Shawl, Abdul. S. and Tantry, Mudasir. A. (2012). "Determination and seasonal variation of chemical constituents of essential oil of *Hyssopus officinalis* growing in Kashmir Valle as incorporated species of Western Himalaya". *Chemistry of Natural Compounds*, *48*, 502-505. doi: 10.1007/s10600-012-0290-5.

Kintzios, Spiridon. E. (2000). *The sage. The genus salvia.* Harwood Academic Publisher, Amsterdam.

Kumar, Peeyush., Mishra, Sapna., Malik, Anushree. and Santosh, Satya. (2011). "Insecticidal properties of *Mentha* species: A review". *Industrial Crops and Products*, *34*, 802-817. doi: 10.1016/j.indcrop.2011.02.019.

Lakušić, Dmitar., Ristic, Mihailo., Slavkovska, Violeta. and Branislava, Lakušić. (2013). "Seasonal variations in the composition of the essential oils of rosemary (*Rosmarinus officinalis*, Lamiaceae)". *Natural Product Communication*, *8*, 131-134.

Lakušić, Dmitar., Lakušić, Branislava., Ristic, Mihailo., Marčetić, M. and Violeta, Slavkovska. (2014). "Seasonal Variations in the Composition of the Essential Oils of *Lavandula angustifolia* (Lamiacae)". *Natural Product Communication*, *9*, 859-862.

Lane, Alexander., Boecklemann, Astrid., Woronuk, Grant. N., Sarker, Lukman. and Mahmoud, Soheil. S. (2010). "A genomics resource for investigating regulation of essential oil production in *Lavandula angustifolia*". *Planta*, *231*, 835-845. doi: 10.1007/s00425-009-1090-4.

Lange, Bernd Markus., Mahmud, Soheil. Seyed., Wildung, Mark. R., Turner, Glenn. W., Lange, Iris., Baker, Raymond, C., Boydston, Rick. A. and Croteau, Rodney. B. (2011). Improving peppermint essential oil yield and composition by metabolic engineering. Proceedings of the National Academy of Sciences., *108*, 16944-16949. doi:10.1073/pnas.1111558108.

Lange, Bernd. Markus. and Amirhossein, Ahkami. (2013). "Metabolic engineering of plant monoterpenes, sesquiterpenes and diterpenes—current status and future opportunities". *Plant Biotechnology Journal*, *11*, 169-196. doi: 10.1111/pbi.12022.

Lavania, Umesh. C., Srivastava, Sarita., Lavania, Seshu., Basu, Surochita., Nandeesh Kumar, Misra. and Yasuhito, Mukai. (2012). "Autopolyploidy differentially influences body size in plants, but facilitates enhanced accumulation of secondary metabolites, causing increased cytosine methylation". *The Plant Journal*, *71*, 539-49. doi: 10.1111/j.1365-313X.2012.05006.x.

Lira-Medeiros, Catarina Fonseca., Parisod, Christian., Fernandes, Ricardo. Avanci., Mata, Camila. Souza., Cardoso, Monica. Aires, and Paulo. Cavalcanti, Gomes Ferreria. (2010). "Epigenetic variation in mangrove plants occurring in contrasting natural environment". *Plos One*, *5*, e10326. doi: 10.1371/journal.pone.0010326.

Lubbe, Andrea. and Robert, Velpoorte. (2011). "Cultivation of medicinal and aromatic plants for specialty industrial materials". *Industrial Crops and Products*, *34*, 785-801. doi: 10.1016/j.indcrop.2011.01.019.

Mader, Eduard., Lohwasser, Ulrike., Börner, Andreas. and Johannes, Novak. (2010). "Population structures of genebank accessions of *Salvia officinalis* L. (Lamiaceae) revealed by high resolution melting analysis". *Biochemical Systematics and Ecology*, *38*, 178-186. doi: 10.1016/ j.bse.2010.01.001.

Maffei, Massimo. E. (2010). "Sites of synthesis, biochemistry and functional role of plant volatiles". *South African Journal of Botany*, *76*, 612-631. doi: 10.1016 /j.sajb.2010.03.003.

Maffei, Massimo. E., Gertsch, Jürg. and Giovanni, Appendino. (2011). "Plant volatiles: Production, function and pharmacology". *Natural Product Reports*, *28*, 1359. doi: 10.1039/c1np00021g.

Mendez-Tovar, Inés., Sponza, Silvia., Manzanera, Asensio. S. Carmen, Schmiderer. Corinna. and Johannes, Novak. (2015). "Volatile fractions differences for Lamiaceae species using different extraction methodologies". *Journal of Essential Oil Research*, *7*, 497-505. doi: 10. 1080/10412905.2015.1064039.

Mendoza-Poudereux, Isabel., Muñoz-Bertomeu, Jésus., Navarro, Alicia., Arrillaga, Isabel. and Juan, Segura. (2014). "Enhanced levels of S-linalool by metabolic engineering of the terpenoid pathway in spike lavender leaves". *Metabolic Engineering*, *32*, 136-144. doi: 10.1016/ j.ymben.2014.03.003.

Mendona-Poudereux, Isabel., Kutzner, Erika., Huber, Claudia., Segura, Juan., Eisenreich, Wolfgana., and Isabel, Arrillaga. (2015). "Metabolic cross-talk between pathways of terpenoid backbone biosynthesis in spike lavender". *Plant Physiology and Biochemistry*, *95*, 113-120. doi: 10.1016/ j.plaphy.2015.07.029.

Miguel, Maria. G., Duarte, Fernanda., Venáncio, Florência. and Regina, Tavares. (2004). "Variation in the main components of the essential oils from the leaves and flowers of Portuguese *Thymus albicans* over a single season". *Journal of Essential Oil Research*, *16*, 169-171. doi: 10.1080/ 10412905.2004.9698685.

Milos, M., Radonic, A., Bezic, N. and Dunkic, V. (2001). "Localities and seasonal variations in the chemical composition of essential oils of *Satureja montana* L. and *S. cuneifolia* Ten". *Flavour and Fragance Journal*, *16*, 157-160. doi: 10.1002/ffj.965.

Miroddi, Marco., Navarra, Michele., Quattropani, Maria. C., Calapai, Fabrizio., Gangemi, Sebastiano. and Gioacchino, Calapai. (2014). Systematic review of clinical trials assessing pharmacological properties of *Salvia* species on memory, cognitive impairment and Alzheimer's disease. *Neuroscience and Therapeutics*, *20*, 485-495. doi: 10.1111/cns.12270.

Mirjalili, Mohammad. Hossein, Saleli., Peyman, Sonboli. Ali. and Masood Mohammadi, Vara. (2006). "Essential oil variation of *Salvia officinalis* aerial parts its phenological cycle". *Chemistry of Natural Compounds*, *42*, 19-13. doi: 10.1007/s10600-006-0027-4.

Mirjalili, Mohammad. Hossein, Tabatabaei, S.M.F., Hadain, J., Nejad Hebraimi. and Sonboli, A. (2007). "Phenological variation of the essential oil of *Artemisia scoparia* Waldst. Kit from Iran". *Journal of Essential Oil Research*, *19*, 326-329. doi: 10.1080/10412905.2007.9699294.

Mirjana, S. and Bezic, Nada. (2004). "Chemical composition and antimicrobial variability of *Satureja montana* L. essential oils produced during ontogenesis". *Journal of Essential Oil Research*, *16*, 387-391. doi: 10.1080/10412905.2004.9698751.

Moghaddam, Mohammad., Pirbalaouti, Ghasemi Abdollah., Mehdiza, Leila., Mohammad, Reza Pirmoradi. (2015). "Changes in composition and essential oil yield of *Ocimum ciliatum* at different phenological stages". *European Food Research and Technology*, *240*, 199-204. doi: 10.1007/ s00217-014-2320-y.

Mukherjee, Pulok. K., Maity, Naladri., Nema, Neelesh. K., Birendra, K. and Sarkar, B. K. (2011). "Bioactive compounds from natural resources against skin aging". *Phytomedicine*, *19*, 64-73. doi: 10.1016/ j.phymed. 2011.10.003.

Muñoz-Bertomeu, Jesús., Arrigalla, Isabel. and Juan, Segura. (2007). "Essential oil variation within and among natural populations of *Lavandula latifolia* and its relation to their ecological areas". *Biochemical Systematics and Ecology*, *34*, 479-488. doi: 10.1016/j.bse.2007.03.006.

Muñoz-Bertomeu, Jesús., Ros, R., Arrillaga, Isabel. and Juan, Segura. (2008). "Expression of spearmint limonene synthase in transgenic spike lavender results in an altered monoterpene composition in developing leaves". *Metabolic Engineering*, *10*, 166–177. doi: 10.1016/j.ymben.2008.04.002.

Naghdi Badi, Hassanali., Yazdani, Darab., Mohammad, Ali. Sajed. and Fatemeh, Nazari. (2004). "Effects of spacing and harvesting time on herbage yield and quality/quantity of oil in thyme, *Thymus vulgaris* L". *Industrial Crops and Products*, *19*, 231-236. doi:10.1016/ j.indcrop. 2003.10.005.

Ohlsson, Ana. B., Segerfeldt, Patrik., Lindströn, Anders., Borg-Karlson, Anna-Karin. and Torkel, Berglund. (2013). "UV-B exposure of indoor-grown *Picea abies* seedlings causes an epigenetic effect and selective emission of terpenes". *Zeitschrift für Naturforschung C*, *68*, 139-147. doi: 10.5560/ZNC.2013.68c0139.

Ormeño, Elena., Fernandez, Catherine. and Jean-Philippe, Mévy. (2007). "Plant coexistence alters terpene emission and content of Mediterranean species". *Phytochemistry*, *68*, 840-852. doi: 10.1016/j.phytochem.2006.11.033.

Ormeño, Elena., Olivier, Romain., Jean-Philippe, Mévy, J. P., Baldy, Virginie. and Catherine Fernandez, C. (2009). "Compost may affect volatile and semi-volatile plant emissions through nitrogen supply and chlorophyll fluorescence". *Chemosphere*, *77*, 94-104. doi: 10.1016/ j.chemosphere. 2009.05.014.

Palá-Paúl, Jesús., Pérez-Alonso, Maria José., Velasco-Negueruela, Arturo., Varadé, Jezabel, Villa., Ana, Maria., Sanz, Jesús. and Brophy, Joseph. J. "Analysis of the essential oil composition from the different parts of *Eryngium glaciale* Boiss. from Spain". *Journal of Chromatography A*, *1094*, 179-182. doi: 10.1016/j.chroma.2005.09.029.

Palá-Paúl, Jesús., Usano-Alemany, Jaime., Soria, Cristina., Pérez-Alonso, Maria José. and Brophy, Joseph. J. (2008). "Essential oil composition of *Eryngium campestre* L. growing in different soil types. A preliminary study". *Natural Product Communications*, *3*, 1221-1126.

Palá-Paúl, Jesús., Usano-Alemany, Jaime., Pérez-Alonso, Maria José. Brophy, Joseph. J. and Cristina, Soria. (2010). "Essential oil composition of the different parts of *Eryngium aquifolium* from Spain". *Natural Product Communications*, *5*, 817-821.

Pandey, Vineeta., Verma, Ram. S., Amit, Chauhan. and Rakesh, Tiwari. (2014). "Compositional variation in the leaf, flower and stem essential oils of Hyssop. (*Hyssopus officinalis* L.) from Western-Himalaya". *Journal of Herbal Medicine*, *4*, 89-95. doi:10.1016/j.hermed.2013.12.001.

Peixoto, Magna. Galvão., Bacci, Leandro., Blank, Arie. Fitzgerland., Albano Araujo, Ana. Paula., Santos Silva, Jefferson. Henrique., Almeida Santos, Abraão., Passos Oliveira, Alexandre., Santos da Costa, Andrea. and Maria de Fátima Arrigoni-Blank. (2015). "Toxicity and repellency of essential oils of *Lipia alba* chemotypes and their major monoterpenes agains stores grain insects". *Industrial Crops and Products*, *71*, 31-36. doi:10.1016/j.indcrop.2015.03.084.

Perry, Nigel. B., Anderson, Rosemary. E., Brennan, Nerida. J., Douglas, Malcom. H., Heaney, Anna. J., McGimpsey, Jennifer. A. and Bruce, M. Smallfield. (1999). Essential oils from dalmatian sage (*Salvia officinalis* L.): variations among individuals, plant parts, seasons, and sites. *Journal of Agricultural and Food Chemistry*, *47*, 2048-2054.

Pichersky, Eran. and Jonathan, Gershenzon. (2002). "The formation and function of plant volatiles: perfumes for pollinator attraction and defense". *Current Opinion in Plant Biology*, *5*, 237-43. doi:10.1016/S1369-5266 (02)00251-0.

Porres-Martínez, María., González-Burgos, Elena., Carretero, Emilia. M. and Gómez-Serranillos, M. Pilar. (2014). "Influence of phenological stage on chemical composition and antioxidant activity of *Salvia lavandulifolia* Vahl. essential oils". *Industrial Crops and Products*, *53*, 71-77.

Prakash, Branu., Kedia, Akash., Prashant, Kumar. Mishra. and Dubey. N. K. (2015). "Plant essential oils as food preservatives to control moulds, mycotoxin contamination and oxidative deterioration of agri-food commodities - Potentials and challenges". *Food Control*, *47*, 381-391. doi: 10.1016/j.foodcont.2014.07.023.

Rajabi, Zahra., Ebrahimi, Mohsen., Farajpour, Mostafa., Mirza, Mehdi. and Hossein, Ramshini. (2014). "Compositions and yield variation of essential oils among and within nine *Salvia* species from various areas of Iran". *Industrial Crops and Products*, *61*, 233-239. doi:10.1016/ j.indcrop. 2014.06.038.

Raut, Jayant Shankar. and Sankuny Mohan, Karuppayil. (2014). A status review of the medicinal properties of the essential oils". *Industrial Crops and Products*, *62*, 250:264. http://dx.doi.org/10.1016/j.indcrop.2014.05.055.

Rios-Estepa, Rigoberto., Lange, Iris., Lee, James. M. and Lange, B. Markus. (2010). "Mathematical modeling-guided evaluation of biochemical, developmental, environmental, and genotypic determinants of Essential oil composition and Yield in peppermint Leaves". *Plant Physiology*, *152*, 2105-2119. doi: 10.1104/pp.109.152256.

Rivera, Diego., Allkind, Robert., Obón, Concepción., Alcaraz, Francisco., Velpoorte, Rob. and Michael, Heinrich. (2014). What is in a name?."The need for accurate scientific nomenclature for plants". *Journal of Ethnopharmacology*, *152*, 393-402. http:// dx.doi.org/ 10.1016/ j.jep. 2013.12.022.

Rodrigues, Leandra., Póvoa, Orlanda., Teixeira, Generosa., Figueiredo, Ana. Cristina, Moldão. Martins. and Ana, Monteiro. (2013). "Trichomes micromorphology and essential oil variation at different developmental stages of cultivated and wild growing *Mentha pulegium* L. populations from Portugal". *Industrial Crops and Products*, *43*, 692-700. doi: 10.1016/j.indcrop.2012.07.061.

Rohloff, Jens., Dragland, Steinar., Mordal, Ruth. and Tor-Henning, Iversen. (2005). "Effect of harvest time and drying method on biomass production, essential oil yield, and quality of peppermint (Mentha x piperita L.)". *Journal of Agricultural and Food Chemistry*, *53*, 4143-4148.

Rzepa, Jósef., Wojtal, Łukasz., Staszek, Dorota., Grygierczyk, Gabriela., Labe, Karina., Hajnos, Michal., Kowalska, Teresa. and Monika, Waksmundzka-Hajnos. (2009). "Fingerprint of selected Salvia species by HS-GC-MS analysis of their volatile fraction". *Journal of Chromatographic Science*, *47*, 575-80. doi: 10.1093/chromsci/47.7.575.

Salido, Sofía., Altarejos, Joaquín., Nogueras, Manuel., Sánchez, Adolfo. and Pascual, Luque. (2004). "Chemical composition and seasonal variations of spike lavender oil from Southern Spain". *Journal of Essential Oil Research*, *16*, 206-210. doi: 10.1080/10412905.2004.9698698.

Sánchez-Muñoz, B., Arturo, Aguilar., Maria, Isabel., King-Díaz, Beatriz., Rivero, José Fausto. and Blas, Lotina-Hennsen. (2012). "The sesquiterpenes β-caryophyllene and caryophyllene oxide isolated from *Senecio salignus* act as phytogrowth and photosynthesis inhibitors". *Molecules*, *17*, 1437-47. doi: 10.3390/molecules17021437.

Santana, O., Cabrera, R., Giménez, C., González-Coloma, A., Sánchez-Vioque, R., de los Mozos-Pascual, M., Rodríguez-Conde, M. F., Laserna-Ruíz, I., Usano-Alemany, J. and and Herráiz-Peñalver, D. (2012). "Perfil químico y biológico de aceites esenciales de plantas aromáticas de interés agro-industrial en Castilla-La Mancha (España)". *Grasas y Aceites-International Journal of Fats and Oils*, *63*, 214-222. doi: 10.3989/ gya. 129611.

Santos-Gomes, Paula. C. and Manuel, Fernandes. (2001). "Organ- and season-dependent variation in the essential oil composition of Salvia officinalis L. cultivated at two different sites". *Journal of Agricultural and Food Chemistry*, *49*, 2908-2916. doi:10.1021/jf001102b.

Savelev, S., Okello, E., Perry, N. S. L., Wilkins, R. M. and Perry, E. K. (2003). "Synergic and antagonistic interactions of anticholinesterase terpenoids in *Salvia lavandulaefolia* essential oils". *Pharmacology, Biochemistry and Behavior*, *75*, 661-668. doi:10.1016/S0091-3057(03)00125-4.

Saviuc, Crina-Maria., Drumea, Veronica., Olariu, Laura., Chifiriuc, Mariana-Carmen., Bezirtzoglou, Eugenia. and Veronica, Lazar. (2015). "Essential oils with microbicidal and antibiofilm activity". *Current Pharmaceutical Biotechnology*, *15*, 137-151. doi: 10.2174/138920101602150112151549.

Schmiderer, Corinna., Grassi, Paolo., Novak, Johannes. and Chlodwig Franz. (2008). "Diversity of essential oil glands of Spanish sage (*Salvia lavandulifolia* Vahl, Lamiaceae)". *Natural Products Communications*, *3*, 1155-1160.

Schwob, Isabelle., Bessiere, Jean-Marie., Massoti, Véronique. and Viano, Josette. (2004). "Changes in essential oil composition in Saint John's wort (*Hypericum perforatum* L.) aerial parts during its phenological cycle". *Biochemical Systematics and Ecology*, *32*, 735-745. doi:10.1016/ j.bse. 2003.12.005.

Selmar, Dirk. and Mark, Kleinwächter. (2013a). "Stress enhances the synthesis of secondary plant products: the impact of stress-related over-reduction on the accumulation of natural products". *Plant Cell Physiology*, *54*, 817-826. doi: 10.1093/pcp/pct054.

Selmar, Dirk. and Mark, Kleinwächter. (2013b). "Influencing the product quality by deliberately applying drought stress during the cultivation of medicinal plants". *Industrial Crops and Products*, *42*, 558-566. doi:10.1016/ j.indcrop.2012.06.020.

Sellami, Ibtissem Hamrouni., Maamouri, Emma., Chahed, Thouraya., Aidi Wannes., Wissem, Elyes. Kchouk, Mohamed. and Brahim, Marzouk. (2009). "Effect of growth stage on the content and composition of the essential oil and phenolic fraction of sweet marjoram (*Origanum majorana* L.)". *Industrial Crops and Products*, *30*, 395-402. doi:10. 1016/j.indcrop.2009.07.010.

Shahhoseini, Reza., Beyraghdar, Arman., Karimi, Razi. and Mohammad, Taghi Ebadi. (2013). "Essential oil content and composition of lemon verbena (*Lippia citriodora* Kunth.) during different phenological stages". *Journal of Medicinal Plants and By-products*, *2*, 205-208.

Shojaiefar, Samira., Mirlohi, Aghafakhr., Sabzalian, Mohammad. R. and Hamid, Yaghini. H. (2015). "Seed yield and essential oil content of fennel influenced by genetic variation and genotype x year interaction". *Industrial Crops and Products*, *71*, 97-105. doi:10.1016/ j.indcrop. 2015.03.055.

Silva, Eric Keven., de Barros, Fernandez., Reginae, Victória., Vilela, Borges., Soraia, Alvarenga., Botrel, Diego. and Fabiana, Queiroz. Water adsorption in Rosemary essential oils microparticles: kenetics, thermodynamics and storage conditions. *Journal of Food Engineering*, *140*, 39-45. doi:10. 1016/j.jfoodeng.2014.05.003.

Singh, Munnu. and Neha, Guleria. (2013). "Influence of harvesting stage and inorganic and organic fertilizers on yield and oil composition of rosemary (*Rosmarinus officinalis* L.) in a semi-arid tropical climate". *Industrial Crops and Products*, *42*, 37–40. doi:10.1016/j.indcrop.2012.04.054.

Smitha, G. R. and Virendra, Rana. S. (2015). "Variations in essential oil yield, geraniol and geranyl aceate contents in palmarosa (*Cymbopogon martini*, Roxb. Wats. Var. *motia*) influenced by inflorescence development". *Industrial Crops and Products*, *66*, 50-160. doi:10.1016/ j.indcrop. 2014. 12.062.

Su, Yu-Chang., Hsu, Kuang-Ping., Wang, Eugene I. and Ho, Cheng-Lu. (2015). "Chemical composition and anti-mildew activities of essential oils from different parts of *Michelia compressa* var.*formosana*". *Natural Products Communications*, *10*, 665-668.

Teixeira, Bárbara., Marques, António., Ramos, Cristina., Neng, Nuno. R., Nogueira, José M. F., Saraiva, Jorge Alexandre. and Maria Leonor, Nunes. (2013). "Chemical compositon and antibacterial and antioxidant properties of commercial essential oils". *Industrial Crops and Products*, *43*, 587-595. doi:10.1016/j.indcrop.2012.07.069.

Thiem, Barbara., Kikowska, Malgorzata., Kurowska, Anna. and Danuta, Kalemba. (2011). "Essential oil composition of the different parts and in vitro shoot culture of *Eryngium planum* L." *Molecules*, *16*,7115–7124. doi: 10.3390/molecules16087115.

Tibaldi, Giorgio., Fontana, Emanuela. and Silvana, Nicola. (2011). "Growing conditions and postharvest management can affect the essential oil of *Origanum vulgare* L. ssp. *hirtum* (Link) Ietswaart". *Industrial Crops and Products*, *34*, 1516-1522.

Tounsi, Moufida. Saidani, Wannes. Wissem, Aidi. Ouerghemmi, Ines. Msaada, Kamel. Smaoui, Abderrazak. and Marzouk, Brahim. (2011). "Variation in essential oil and fatty acid composition in different organs of cultivated and growing wild *Ruta chalepensis* L." *Industrial Crops and Products*, *33*, 617-623. doi:10.1016/j.indcrop.2010.12.028.

Tsiri, Dimitroula., Kretsi, Ourania., Chinou, Ioanna. B. and Carolinne, Spyropoulos. G. (2003). "Composition of fruit volatiles and annual changes in the volatiles of leaves of *Eucalyptus camaldulensis* Dehn. Growing in Greece". *Flavour and Fragance Journal*, *18*, 244-247. doi: 10.1002/ffj.1220.

Usano-Alemany, Jaime. (2012). "Influencia de los factores ambientales y agronómicos sobre la producción de los aceites esenciales de *Salvia lavandulifolia* Vahl." Ph. Diss. Universidad Complutense de Madrid, Spain. http://eprints.ucm.es/16629/.

Usano-Alemany, Jaime. and David, Herráiz. (2014). *Diversidad fitoquímica de la salvia española. Perfil química y biológico de los aceites esenciales de la especie Salvai lavandulifolia Vahl.* Saarbrücken, Deutschland. Publicia.

Usano-Alemany, Jaime., Palá-Paúl, Jesús., Santa-Cruz, Rodríguez., Manuel, and David, Herraiz-Peñalver. (2014a). "Chemical description and Essential oil yield variability of different accessions of *Salvia lavandulifolia* Vahl". *Natural Product Communications*, *9*, 273-276.

Usano-Alemany, Jaime., Palá-Paúl, Jésus. and David, Herráiz-Peñalver. (2014b). "Comprehensive phenological description of essential oil chemotypes of *Salvia lavandulifolia* Vahl. grown under the same environmental conditions". *Chemistry and Biodiversity, 11*, 1963-1977. doi: 10. 1002/cbdv.201400090.

Usano-Alemany, Jaime., Palá-Pául, Jesús. and David, Herráiz-Peñalver. "Essential oil yields and qualities of different clonal lines of *Salvia lavandulifolia* monitored in Spain over four years of cultivation". *Industrial Crops and Products* in press. doi:10.1016/ j.indcrop. 2015.1 1.010.

Vanzani, Paola., Rosseto, Monica., De Marco, Veronica., Sacchetti, Linda. E., Paoletti, Maurizio. G. and Adelio, Rigo. (2011). "Wild mediterrean plants as traditional food: a valuable source of antioxidants". *Journal of Food Science, 76*, 46-50. doi: 10.1111/j.1750-3841.2010.01949.x.

Yadav, Ritesh. K., Sangwan, Rajender. S., Sabir, Farzana., Srivastava, Awadesh. K. and Neelam, Sangwan. S. (2014). "Effects of prolonged water stress on specialized secondary metabolites, peltate glandular trichomes, and pathway gene expression in *Artemisia annua* L. plant" *Physiology and Biochemistry, 74*, 70-83. doi: 10.1016/j.plaphy .2013.10.023.

Yosr, Zaouali, Hnia., Chograni, Rim. Trimech. and Boussaid, Mohamed. (2013). "Changes in essential oil composition and phenolic fraction in *Rosmarinus officinalis* L. var. *typicus* Batt. organs during growth and incidence on the antioxidant activity". *Industrial Crops and Products, 43*, 12-419. doi:10.1016/j.indcrop.2012.07.044.

Zouari, Nacim., Iyadi, Imen., Fakhfakh, Nahed. Rebai, Ahmed. and Sami, Zouari. (2012). "Variation of chemical composition of essential oils in wild populations of *Thymus algeriensis* Boiss. et Reut., a North African endemic Species". *Lipids in Health and Disease, 11*, 28. doi: 10. 1186/ 1476-511X-11-28.

In: Essential Oils
Editor: Miranda Peters

ISBN: 978-1-63484-351-5

Chapter 3

EXTRACTION TECHNIQUES AND PHARMACOLOGICAL POTENTIAL OF ESSENTIAL OILS FROM MEDICINAL AND AROMATIC PLANTS OF MAURITIUS

Z. Aumeeruddy-Elalfi and F. Mahomoodally*
Department of Health Sciences, Faculty of Science,
University of Mauritius, Réduit, Mauritius

ABSTRACT

Essential oils (EOs) are natural volatile complex blends of biologically active molecules used nowadays for a panoply of applications. Presently, their importance have been highlighted due to the increasing demand by the food and pharmaceutical industries but also for their potential to provide therapeutic benefits in the prevention and management of diseases. In view of the multiple applications of the EOs, it is becoming important to explore different extraction techniques for higher yields of EOs as well as for the isolation of new biologically active compounds. This chapter attempts to highlight recent extraction techniques for obtaining EOs from medicinal aromatic plants. The conventional extraction techniques, their optimisation and improvements are discussed through the innovative principles, their benefits and disadvantages. Moreover, this chapter also endeavors to provide an up-to-date literature on EOs, the reported traditional usage of EOs from medicinal and aromatic plants of Mauritius, common extraction methods, bioactivities, and applications.

Keywords: essential oils, extraction, traditional, Mauritius

* Corresponding author: Tel: +230 4037578; fax: +230 4656928. E-mail address: f.mahomoodally@uom.ac.mu (M.F.Mahomoodally).

1. INTRODUCTION

Aromatic medicinal herbs and food plants have served as rich sources of biologically active substances (Delmare et al. 2007). Diverse array of purposes have been found for medicinal herbs including medicine, nutrition, beverages, dyeing, repellents, fragrances, cosmetics, charms, smoking and other industrial applications. According to existing literature, herbs are still found in 40% of prescription drugs (Newman and Cragg 2007). Aromatic plants for example those from the Lamiaceae and Umbelliferae family are now cultivated worldwide, mainly for use as culinary and medicinal herbs and are widely studied as natural antioxidant sources since they are enriched in polyphenols (Delamare et al. 2007). Their potent bioactivity and relatively low toxicity have rendered them useful as ingredients in complementary and alternative medicine and as nutritional supplements (Lindeman et al. 2014).

EOs also known as ethereal oils, characterized as volatile, natural, liquid, complex components with a strong aroma (Guenther 1948). These EOs are extracted from aromatic plants, mostly from vegetable organs (e.g., flowers, leaves, barks, woods, roots, rhizomes, fruits and seeds) through different processes such as expression, fermentation, enfleurage or extraction. These oils consist mainly of terpenes and oxygenated compounds, obtained through a form of distillation from aromatic plant materials of a single botanical species and form (Bakkali et al. 2008). The constituents of EOs are produced in the plant through the mevalonic and shikimic acid pathways as secondary metabolites and stocked in the plants in different organs depending on the parts of the plant it is found: glandular trichomes, oil cells or ducts in plant tissues (Chaiyana et al. 2012; Hunter 2009).

Major advances in EO analysis have been feasible with the introduction of powerful instrumentation. Gas Chromatography coupled with multivariate curve resolution method gives a profile which provides a close illustration of the EO (Bansal et al. 2014). However, the procedure for the extraction and isolation of EOs often influences the profile by chemically changing the composition of the EOs (isomerization, saponification, polymerization which result in the formation of artifacts). New methods have been established which aim at extracting without solvent, decreasing the extraction time in order to have unaltered composition of the EO and to limit the formation of artifacts.

2. METHODS OF EXTRACTION

2.1. Supercritical Fluid Extraction Method

The type of distillation is known to change the chemical constituents of EOs. During the last decade, the vast majority production of EOs was obtained by the process of distillation and solvent extraction. The Supercritical Fluid is a method of extraction which is performed mostly with carbon dioxide (CO_2) and is maneuvered at low temperature, which is crucial for the conservation of the biological properties of plant materials (Beckman 2004). CO_2 is the Supercritical Fluid of choice for this method as it has a low critical pressure and temperature (74 bars, 32°C), it is known to be nontoxic (environment friendly), is inflammable and easily removed from the extract at the end of the process. The use of CO_2 is also beneficial as it allows penetration into small pores due to its low surface tension and viscosity (Beckman 2004; Dixon and Johnston 1997). The EOs yielded from this method are of high-quality and optimal for the production of pharmaceutical drugs, to be incorporated as organic additives in the food industries and in cosmetology. Recently, much more emphasis has been set on the extraction of EOs using the Supercritical CO_2 method. Bagheri et al. (2014) reported that Supercritical CO_2 extraction method yield an EO richer in antioxidant activity (EC50 = 103.28 ± 0.05 μg/ml, where EC50 = concentration of test material required to cause a 50% decrease in initial scavenging of 1,1-Diphenyl-2-picryl-hydrazyl (DPPH)) than hydrodistillation method (EC50 = 316.27 ± 0.12 μg/ml) and explained it as due to the thermal degradation, hydrolysis, and solubilization of part of the components in the water used for the extraction by hydrodistillation, which can change their antioxidant capacity. Or due to the fact that the water used in hydrodistillation process renders several antioxidants unstable or breaks them down by enzymatic action in the wet plant material. Xiao et al. (2007) reported that the Maximum global yields were obtained using an extraction temperature of 35°C, dynamic time, 35 minutes; pressure, 15 MPa and modifier volume (for optimum solubility) of 40 mL.

The Supercritical method has also been used for the extraction of EO from functional food. For instance, *Citrus sphaerocarpa Tanaka*, a popular sour citrus in Japan has been extracted using the Supercritical method as reported by Suetsugu et al., (2013). They reached a maximum yield of 1.55% (by weight of wet sample) at the temperature of 353 K and pressure of 20 MPa with this method of extraction, which was 13 times higher than that obtained by the conventional cold-press method. In addition, Suetsugu et al., (2013) also reported that the composition of the EO extracted by the Supercritical method was low in monoterpenes and high in oxygenated compounds, sesquiterpenes, which are known to accentuate the aromatic characteristics of EOs.

However, as in every method, there is a downside. For the Supercritical method, the extractions of polar analytes are not efficient. This limitation has been overcome as reported in the literature, by the addition of small amounts of polar substances, such as methanol or ethanol to the supercritical CO_2, thus improving its solution power.

Nowadays public awareness towards biosafety hazards are gaining in amplitude. The use of solvents is being more and more pinpointed as to the possible residual contaminants left in the final products. Also, the development of new and clean technologies together with the new requirements of the food and medical industries for ultra-pure products give the Supercritical Fluid method increasing importance. Moreover it has been reported previously as a method for the extraction of valuable compounds from natural products (Hartono et al. 2001).

2.2. Short-Path Molecular Distillation

Another recently reported technique is the Short-Path Molecular Distillation method (Tovar et al. 2010; Martins et al. 2012; Olmedo et al. 2014). Short-Path Molecular Distillation, in general, is a special technologically advanced thermal separation technique which functions at high vacuum, and is used in most cases to enhance the concentration of a specific molecule, thus increasing the quality for the EO extracted. For instance, Martins et al., (2012), demonstrated the enhancement of a molecule, methyl chavicol, from the sweet basil by this method. Also, Chen et al., (2012) illustrated the enrichment of two compounds, cuminaldehyde and *p*-mentha-1, 4-dien-7-al in the *Cuminum cyminum* EO by the Short-Path Molecular Distillation.

The Short-Path Molecular Distillation is also known as molecular distillation/ evaporation. This method covers a range of applications not possible in traditional distillation/evaporation units which operates with high temperature and high vacuum. This method is used for producing compounds based on their boiling points (Olmedo et al. 2014). It is also a method of separation and purification of thermally unstable materials with low vapor pressures and high molecular weight, without leading to thermal decomposition. Due to the very short contact time with the heating surface the thermal impact on heat susceptible products is minimal. Thus, when working at comparatively high temperatures requisite to separate some higher boiling compounds, the impact on the product is minimized.

The Short-Path Molecular Distillation is perfect for running different products in a short interval of time, particularly high value products, as with this method there is minimal product loss at the end of the manipulation. Short-Path Molecular Distillation is ideal for applications in the edible oil and EO production, flavor industry, recovery of fatty acids such as omega 3 (Olmedo et al. 2014).

Olmedo et al. (2014), reported that the fractions obtained from oregano EO by Short-Path Molecular Distillation can be used as natural antioxidants for food products sensitive to lipid oxidation as well as additives to be integrated in organic food, since the fractions obtained by this process was reported to exhibit greater antioxidant activity than the whole oregano EO obtained by hydro-distillation in a Clevenger type apparatus.

Tovar et al. (2011) reported that high quality EO can be extracted from Lemongrass by the Short-Path Molecular Distillation method. The outcome of their experiment demonstrated an increase in citral concentration from19.816 $mgmL^{-1}$ (initial sample) to 40.963 $mgmL^{-1}$ (at 120°C and 4.5 $mL\ min^{-1}$) of the bioactive compound (citral) in the distillate stream.

Short-Path Molecular Distillation method has also been investigated recently as a method for removing toxic substances from EOs, rendering them more safe for use for example as food additives. For instance, Xiong et al. (2013) reported the significant removal of 3 phthalates compounds in the EO of *Citrus sinensis* (L.), producing a refined oil without destroying the perfume. The clearance rate of phthalates compounds (contaminants present in some EOs and which at a high dose are considered as health hazards) was found to be varied according to different evaporator conditions such as temperature, pressure and flow feed rate. The optimal conditions for the removal of phthalates from the EO by molecular distillation were calculated at an evaporation temperature of 50 °C, evaporator pressure of 5 kPa and a feed flow rate of 0.75 ml/min.

2.3. Microwave Assisted Extraction

Another alternative method which has been suggested for extracting natural products is the microwave assisted extraction method. Different method using the microwave as pretreatment or incorporated in a unique process are available and has been reported recently in the literature (Uysal et al. 2010; Golmakani et al. 2008). Microwave extraction is now widely accepted as an efficient technique that significantly lessens extraction time while increasing the yield and the quality of the extract (Abert Vian et al. 2008; Ferhat et al. 2008). Okoh et al., (2010) demonstrated that when comparing the EOs of the leaves of *Laurus nobilis* L. obtained by hydrodistillation and that extracted by the solvent free microwave assisted method, the strongest antioxidant activity against DPPH radical was found in the EO obtained by the solvent free microwave assisted method. Amirah et al., (2012) demonstrated that the Microwave Assisted Extraction was more favorable compared to Ultrasonic Assisted Extraction for the extraction of stem bark of *Jatropha curcas.* The gallic acid yielded in this experiment by these two methods was more significant when using the microwave method rather than the ultrasonic method.

Microwave-mediated processes offer a number of advantages in EO extraction. For instance, the equipment size is reduced compared to other methods, and it's a technique which is easy to use, fast, solvent free, all of which contribute to reduce environmental impact and costs. Recent procedures based on microwave extraction have substituted some of the conventional processes and other thermal extraction techniques that have been used more commonly for decades.

2.3.1. In Situ Microwave-Generated Hydrodistillation

In this method, no solvent is required to extract the EO. The microwave acts on the water molecules contained in the cells of the plant material, whereby molecular motion within polar or ionic components causes heating, leading to the rupture of the glands and oleiferous receptacles, containing the EOs. The pure EO is collected at the end of this process without any preliminary clean up. The advantage of this method is that it is green, rapid (30 minutes), there is an increase in yield (Ferhat et al. 2007; Da Porto et al. 2009), no artifacts and oil components rearrangements formed as the plant material are not soaked in a liquid which may result in the formation of artifacts and there is no variation in pH in the plant material (which is the case when the plant material are soaked in an aqueous medium and in this condition, the EO components are prone to experience rearrangements). The composition of the EO varies with the different methods of extraction. The volatile secondary metabolites found in the EOs vary in abundance with the variation in extraction method. For instance, Binello et al., (2014), reported high yield in coumarin (16.59%), 7-methoxycoumarin (11.81%) and caryophyllene oxide (4.26%) for the EO *Lavandula angustifolia* Mill. extracted by the *in situ* microwave generated hydrodistillation process. When the composition of *Origanum vulgare* L. obtained by the *in situ* microwave generated hydrodistillation process was compared to the EO composition of the traditional hydrodistillation process, carvacrol was found to be the most abundant in the *in situ* method compared to that in the hydrodistillation one (35.01% and 26.70% respectively), and the quantity of 4-terpineol was reported significantly more abundant in the *in situ* method (18.05%) than in hydrodistillation method (7.44%).

On the other hand, in a recent example, Orio et al. (2013) established that there was no significant difference in EO yield and composition of the three mint species (*Mentha spicata* L. var. *rubra*, *Mentha spicata* L. var. *viridis* and *Mentha piperita* L.) tested, when the *in situ* microwave generated hydrodistillation method was compared to the reference method in EO extraction that is the hydrodistillation method. The composition in the most abundant compound was reported to be different for the three species and there was a slight increase in the concentration of the most abundant compounds. For instance, for *M. spicata* L. var. *rubra* carvone, limonene and β caryophyllene were the most abundant compounds observed for the *in situ* and the hydrodistillation methods when assessed using both dry and fresh extracted plant materials.

2.3.2. Microwave Steam Diffusion

Microwave Steam Diffusion is another novel method for green extraction of EOs of aromatic plants. The concept of this method is the association of a heating apparatus of microwave energy to another body which provides saturated steam. The steam, when in contact with the plant material in the microwave heating reactor, favors the liberation of the EO from the plant tissues. The steam thus causes the diffusion of the EO towards the cooling system. This method favors many advantages in terms of energy saving, low cost, short extraction period and also offers strong penetration force as well as high selectivity (Ferhat et al. 2011). Further studies by Ferhat et al., (2011), demonstrated that the EOs extracted from orange peel by the Microwave Steam Diffusion apparatus, with extraction time 12 minutes resulted with similar yield and chemical composition as that obtained by conventional steam diffusion method for an extraction time of 40 minutes. A similar system to this one is the Microwave Steam Distillation, described by Navarrete et al., (2011); Sahraoui et al., (2008,). In the latter, the system is comparable to that of the conventional steam distillation but with the addition of a microwave irradiation pretreatment incorporated to break the EO containing glands thus facilitating extraction and giving a better yield in a shorter time frame.

On the other side, Périno-Issartier et al., (2013) demonstrated that the yield of *Lavandula Grosso* extracted by the Microwave Steam Diffusion method was significant and comparable to the yield of the same plant material extracted by the *In situ* microwave-generated hydrodistillation. The Microwave Steam Diffusion method has been noted in the literature as a method which provides the enrichment in selective aromatic compounds and in so doing, increasing the numerous activities that these aromatic compounds can confer to the EOs such as an increase in their antibacterial, antioxidant and anti-inflammatory properties.

2.4. Ultrasound Assisted Extraction

This method applies the ultrasound expertise to the extraction method. The mechanical stress exerted through the ultrasound induces cavitations in the plant material which result in cellular breakdown. Ultrasound assisted extraction is considered in recent literatures as an effective method for the extraction of active compounds from plant materials (Wang et al. 2011; Musato et al. 2011). It is not a green method like the microwave assisted methods but with this method, extraction time is lessened and the yield is improved. This method showed superior extraction effectiveness compared to steam distillation or superheated water extraction. Jayanthi and Lalitha, (2013) investigated on the comparison of two methods: the conventional reflux method and the ultra sound assisted extraction method for the extraction of *Eichhornia crassipes*. It was reported that the extraction of fresh material was more efficient by the ultrasonic method which generated a yield of 1330 mg compared to 590 mg with the reflux method using water as the solvent of extraction. However, the extraction of dry plant material gave more significant result with the reflux method which was accounted by the fact that in both methods there are different variable factors. For instance, in the reflux method, there can be a rapid raise in temperature and condensation which facilitate the liberation of EO in the extraction solvent. Also, there is a difference between fresh plant material and dry plant material, whereby in the latter, there are postharvest alteration of plant metabolites which has to be taken into consideration.

Ultrasound assisted extraction has also been adapted with extraction using supercritical CO_2. When comparing the normal extraction with supercritical CO_2 method, the former was reported by Yu-Chiao et al., (2014) to give better extraction yield for the extraction of *Syzygium aromaticum* EO (13.5% increase). The ultrasound assisted extraction using supercritical CO_2 was concluded by Yu-Chiao et al. (2014) to be an efficient method of EO extraction which is established using gentle extraction conditions (temperature, pressure and CO_2 flow rate.).

Recent research based on ultrasound assisted extraction is the use of ionic liquids for the extraction of plant material at room temperature. Ionic liquids are applicable to EOs as these liquids allow the dissolution of non-polar compounds, are biodegradable, allow the enrichment of extracts and are thermally stable (Zu et al. 2012). Ionic liquids are also used as alternative to Supercritical Fluid and conventional organic solvent and have proved in the literature to be a noteworthy method of extraction applicable for the extraction of medicinal products (Bi et al. 2011).

Da porto et al. (2009) alternatively investigated on the comparison of the extraction of different species of *Mentha spicata* by the Ultrasound-assisted extraction method coupled to a vacuum distillation, with that of the conventional hydrodistillation process. The outcomes have demonstrated that the ultrasonically assisted extraction method coupled with vacuum distillation produced a yield which is more significant than the yield obtained by the hydrodistillation process. Also, it was put forward that the EO produced by the former method was higher in flavoring strength due to the increased concentration of desirable oxygenated compounds (from 5 to 8 times) compared with that of the EO produced by the hydrodistillation.

Wang et al. (2011) worked on probing the use of ultrasound-assisted extraction to extract three natural compounds, including tracheloside, hemislienoside, and arctiin from *Hemistepta lyrata* and monitoring whether the different factors such as extraction solvent, solvent concentration, solvent to material ratio, and extraction time have an effect directly or indirectly on the extraction. The results acquired illustrated that the optimal parameters to extract the target compounds from *H. lyrata* were achieved using an extraction solvent consisting of a mixture of 70% aqueous ethanol having a solvent to material ratio of 20:1 (v/w, ml/g) and with an extraction time of 20 minutes at an ultrasonic frequency of 40 Hz and an extraction temperature of 30 °C. With these optimal conditions of extraction, the Ultrasound assisted extraction was suggested as an effective method for wider applications in the extraction of medicinal plants.

3. Characterization

EO analysis is based mostly on separation techniques giving the best performance, achieved by the most effective tool. The most popular tool used by scientists for separation techniques is the Chromatography and coupled to that is often the Mass Spectrometry for the identification of components. Analysis of EOs have recntly known major developments with varying methods adapted from the conventional Gas-Chromatography coupled to mass spectrometry technique. The driving force of this surge has been the characterization and identification of the structure of known and novel molecules. The advantage of using a Gas chromatograph is that it provides the conditions required for achieving the separation of analyte components without lowering the performance of the column. During the last decades, the Gas chromatography methods have known many advances to improve resolution, enhance performance and sensitivity. Among the most common, we have heard of the fast Gas Chromatography coupled to Mass Spectrometry, the Gas Chromatography coupled in tandem to another Mass Spectrometry, the Gas Chromatography-Mass Spectrometry coupled with the multivariate curve resolution.

3.1. Fast Gas Chromatography-Mass Spectrometry

Fast Gas-Chromatography method exists since the 1960's (Desty et al. 1962) but it is only recently that this method has been revered and used (Rubiolo et al. 2008). Time analysis is a variable factor in the characterization of EO. In Fast Gas-Chromatography method, the aim is to trim down analysistimes by reducing the internal column diameter (as shorter analysis times tend to amplify sample throughput) without diminishing the resolution necessary to effectively determine the peaks of interest and in the process sorting out the specific elution patterns. In order to keep the standard of the Gas Chromatography method, that is not to overcome the resolution problem which may arise when the column is shorten, the oven temperature ramp rates as well as the linear velocity of the carrier gas are increased. A few modifications are put together such as narrowing the internal diameter of the column, using Hydrogen instead of Helium as carrier gas and using a low film thickness. Mondelo et al. (2004) reported the success in demonstrating the effectiveness of the fast Gas Chromatography for the characterization of the five citrus EOs (bergamot, bitter orange, sweet orange, mandarin, and lemon). The Fast Gas Chromatography method was compared to the conventional method revealing a radical decrease in analysis time (around 3 minutes instead of 47) without alteration of the resolution. The technique was defined as reliable in terms of rapid and accurate identification.

Moreover, fast gas chromatography method is also sometimes combined with a high speed quadrupole mass spectrometer, which is capable of functioning under high speed gas chromatography conditions. The advantage of the quadrupole mass spectrometer is that it is multitasking and can proceed to the scanning mode and give multiple reaction monitoring data all together without delay.

Recently, Tranchida et al. (2013) demonstrated the possibility of analyzing qualitatively untargeted compounds present in the Citrus EO as well as the qualitative and quantitative analysis of targeted compounds namely: o-phenylphenol, butylated hydroxytoluene, butylated hydroxy-anisole, with significant accuracy.

3.2. Gas Chromatography Coupled to Tandem Mass Spectrometry

When it comes to more complex analysis, Gas Chromatography can be insufficient or difficult to interpret. Presently, we have seen in the literature the use of the Gas Chromatography coupled to Tandem Mass Spectrometry. It is a powerful analytical technique which offers the possibility of detecting specific, targeted compounds whether present in large amount or in trace.

Following the separation by Gas Chromatography, the tandem mass spectrometry operates by selecting the target ions having specific and known mass. These ions are then dislocated by collision with Helium molecules. The product ion resulting from this collision gives a spectrum which confirms the target analyte as even if there is another ion with the same mass, the spectrum will be different. This factor increases the selectivity of the tandem mass spectrometry. The target gas, which can be Argon, Xenon, Helium or other (according to choice of energy desired for the collision ion dissociation process), can play an important role in the results as the pressure and temperature of the target gas affect the internal energy distribution and thus affecting also the mass spectrum. Hence, low energy target gas is less reproducible. Whereas the high energy target gas for the collision ion dissociation process was found to be more reproducible and to give less rearrangement in the mass spectrum making it less complex to analyze (Raina and Hall 2008; Marriott et al. 2001). According to the literature, Gas Chromatography coupled to Tandem Mass Spectrometry is not commonly use for the regular analysis of EOs but it remains however an accurate tool for the separation and detection of trace elements found in a complex mixture.

3.3. Gas Chromatography Coupled with Multivariate Curve Resolution

Latest trend in Gas Chromatography-Mass Spectrometry analysis in view of improving analysis of EOs have recently been leading to the multivariate curve resolution method. The Gas chromatography technique has many advantages but there are also a few disadvantages, some of them can sometimes be sorted out. Some of the most common issues are the variation of baseline, changes in peak shape and co-elution of compounds having same retention time values, same mass, same boiling point, which give rise to mismatch or low similarity matches in the Mass spectrometry library when analyzing the mass spectrum and identifying the components of EOs. The use of Gas Chromatography-Mass Spectrometry coupled with the multivariate curve resolution techniques is often the key to the co-elution hitch of volatile components. This technique brings high accuracy and reliability to results obtained for the Chromatography-Mass Spectrometry tool. Pakniyat and Mousavi, (2014) reported an increase in the number of identified volatile components from 37 to 80 in the EO of Ziziphora *tenuior* through this method. The Multivariate Resolution method lead to pure chromatogram and mass spectra which can be used for qualitative and quantitative analysis through the Mass Spectrometry library and Kovats retention index for the qualitative analysis and Peak integration for quantitative analysis. This method has been reported to be a possible alternative to the Gas Chromatography coupled to tandem Mass Spectrometry depending on the nature of the analysis.

4. Isolation

The next step after characterization is often the isolation of molecules identified by the analytical tool such as the Gas Chromatography-Mass Spectrometry. One of the interesting tool pin pointed in the literature recently is the Preparative Gas Chromatography. The Preparative Gas Chromatography has revealed to be interesting for the isolation of volatile components. For instance, many natural compounds have been isolated from their EO by this process. Yang et al. (2011) isolated five volatile components, namely the β-Elemene, curzerene, curzerenone, curcumenol and curcumenone from the extract of Curcuma rhizome. Park et al. (2012) isolated the molecules of menthol and menthone from peppermint oil and Zuo et al. (2012) isolated the cis- and trans-Asarone molecules from the EO of *Acorus tatarinowii.*

EOs of many plants, which are valuable natural products with multiple pharmacological activities and rich in volatile components, are very likely to be effectively isolated by the Preparative Gas Chromatography which represents a significant and practical alternative to other preparative techniques such as the preparative high-performance liquid, particularly for volatile compounds. Gas Chromatography, hyphenated to other methods is the popular tool being used nowadays to cut down on the isolation procedures.

The Preparative Gas Chromatography is being revered more and more for the isolation of natural products, particularly terpenes and other volatiles. Ozek and Demirci (2013), has appraised with a full description of the isolation of method of natural products by the Preparative Gas Chromatography, starting from the classification of gas chromatography separations, columns, and stationary phases, sampling techniques, mobile phases, detectors, and fraction collector, to the systems of analytical and preparative gas chromatographic separations.

5. Reported Uses of EOs from Medicinal and Aromatic Plants of Mauritius

Tropical islands like Mauritius are endowed with rich floral biodiversity having an interesting microcosm with diverse species including aromatic medicinal plants, offering interesting medicinal properties. In Mauritius, EOs are highly used in aromatherapy which is an important market well valued in the tourism industry. However, the traditional use of aromatic plants of Mauritius has been long used as part of the traditional medicine for preventive measures against several diseases but also to treat diseases and pains. Table 1 summarizes the documented medicinal and aromatic plants that are used by the local people and which are also well known in folk medicine.

6. Biological Properties of EOs

EOs and extracts of aromatic plants have been utilised and recognized for many years as useful source of pharmaceutical agents and food additives (Juliano et al. 2000). Much more importance have been given to EOs from different plant species these last decades due to the rising interest in their antioxidant, antibacterial and insecticidal properties (Burt 2004). These volatile oils have also functions in chemical defense, as insecticides, acaricides, avoiding fungi phytopathogen colonization, attracting natural enemies of herbivores (Yadav et al. 2008; Karamanoli et al. 2005) and numerous pharmacological properties.

Biologically active compounds of EOs have been documented to possess numerous medicinal properties for the pharmaceutical and nutraceutical industries. The major compositions of EOs and their biological properties are summarized in Table 2.

Table 1. Medicinal and aromatic plants of Mauritius and their reported properties

Scientific Name	Common Name	Utilised parts	EO Properties	Reference
Achyranthes aspera	Herbe sergent.	whole plant.	Antifungal	Gurib-Fakim, A et al.1995.
Acorus calamus	Acore odorant	whole plant.	Antibacterial against *M. tuberculosis.*	Gurib-Fakim, A et al.1995.
Adansonia digita	Baobab	Fruits, Leaves, Seeds	Antimicrobial and anti-inflammatory	Gurib-Fakim, A., Brendler, 2004.
Aegle marmelos	Bael, Bengal Quince	Fruits, roots	Exhibit toxicity to fungi.	Gurib-Fakim, A., Brendler, 2004.
Aframomum melegueta	Alligator pepper	Seeds	Active against *S. aureus*, *B. subtilis*, *P. aeruginosa.C. albicans*, *A. niger*,	Brendler et al. 2010.
Ageratum conyzoides	Herbe de bouc	Aerial parts	Antifungal. Antibacterial, analgesic in mice.	Gurib-Fakim, A., Brendler, 2004.
Apium graveolens	Céleri.	Whole plant.	Anti-inflammatory, tonic, hypoglycemic	Gurib-Fakim, A et l.1995.
Aloysia triphylla	Verveine de France, Lemon verbena	Leaves	Insectidal. Moderate antibacterial action	Gurib-Fakim, A., Brendler, 2004.
Ambrosia tenuifolia	Armoise, Absinthe, Herbe Solférino, Prennial Ragweed.	Leaves, Roots, Stem.	Antibacterial against *S. typhii*, *E. coli*, *P. vulgaris*, *S. aureus*, *B. subtilis* and antifungal against *C. utilis*, *A. niger*, *C. albicans.*	Gurib-Fakim, A et l.1995.
Argyreia nervosa	Liane d'argent, Elephant climber	Leaves, Seed,Stem	EO present antiseptic activity against Gram negative bacteria and phytopathogenic fungi.	Gurib-Fakim, A., Brendler, 2004.
Artemisia absinthium	Absinthe, Armoise	Leaves	Antibacterial, antifungal, antimalarial	Gurib-Fakim, A., Brendler, 2004.
Artemisia afra	African wormwood, armoise d'Afrique	Leaves, twigs.	Antimicrobial, antimalarial, antioxidant, antinematodal.	Brendler et al. 2010.
Artemisia verlotiorum	Brède chinois, Armoise	Leaves	Pullicidal, antibacterial, antifungal.	Gurib-Fakim, A., Brendler, 2004.

Table 1. (Continued)

Scientific Name	Common Name	Utilised parts	EO Properties	Reference
Artocarpus heterophyllus	Jackfruit, jacquier.	Wood, bark, roots, young fruits, leaves	Antioxidant, antihypertensive, antibacterial	Gurib-Fakim, A., Brendler, 2004,132
Ayapana triplinervis	Ayapana, yapanna.	Leaves.	Used against epigastric pain, nausea, vomitting, flatulence.	Gurib-Fakim, A., Brendler, 2004.
Azadirachta indica	Neem	whole plant.	EOs obtained from the seeds exhibit antibacterial properties against gram negative and gram positive bacteria, helps in ulcer treatment, and helps against arthritis. EOs obtained from flowers showed the presence of Azadirachtin and dipropyldisulfide which are insecticidal.	Gurib-Fakim, A et l.1996.
Boswellia sacra	Mohor,Shabi, Shazri, Luban.	Resin	Anti-inflammatory, analgesic, antifungal on *B. carteri* and *B. papyrifera*, antiseptic, astringent, carminative	Brendler et al. 2010.
Calendula officinalis	Souci, Pot Marigold, Clendula	Flowers, Leaves, seeds.	EOs of flowers are cytotoxic, anti tumor promoting EO of leaves are anti-inflammatory and antibacterial against *E. coli*, *S. faecalis*, *S. aureus*, *P. aeruginosa.*	Gurib-Fakim, A et l.1995.
Calophyllum inophyllum	Takamaka	Roots, bark, leaves, resin.	Insecticidal, anti-inflammatory, antiviral.	Gurib-Fakim, A., Brendler, 2004.
Canaga odorata	Ylang ylang, Langilang.	Flowers	Used for its aromatic properties.	Gurib-Fakim, A., Brendler, 2004,.
Canarium madagascariensis	Ramy, Tambazotse, Mana.	Resin	EOs extracted from the resin are used for dental caries and rheumatism	Gurib-Fakim, A., Brendler, 2004.
Canarium paniculatum	Bois colophane.	Leaves, resin.	EOs extracted from the resin have anti-inflammatory properties and reduce joint swelling.	Gurib-Fakim, A., Brendler, 2004.

Scientific Name	Common Name	Utilised parts	EO Properties	Reference
Caprifoliaceae	Thunb.	Flowers, Leaves, Stem.	Bactericidal, antiviral	Gurib-Fakim, A et l.1995.
Cardiospemum halicacabum	Poc poc sauvage, Bonnet de pretre	Roots, leaves, whole plant.	Anti-inflammatory, antibacterial.	Gurib-Fakim, A., Brendler, 2004.
Carthamus tinctorius	Safran batard, Jafron, Safflower.	FLowers	Antibacterial against *S. aureus*, *B. subtilis*, *E.coli*, *S. typhii*, *V. cholera*, antifungal against *C. albicans*, *T. rubrum*, *Trichophyton mentagrophytes*, *Candida albicans.*	Gurib-Fakim, A et l.1995.
Carissa edulis.	carissa	Leaves, root bark.	antiviral	Brendler et al. 2010.
Cedrelopsis grevei	Bois de catafaye, katrafay, Katafa.	Leaves, Bark.	EO extracted from leaves are known as fixative flavours and to have antibacterial properties.	Gurib-Fakim, A., Brendler, 2004.
Chenopodium ambrosioides.	Herbe botrice, Bautrice, Thé du Mexique, Herbe.	Leaves, Stem	Antifungal against *Trichophyton rubrum*, *Microsporum gypseum and* Alntibacterial against *S. aureus and P. aeruginosa.*	Gurib-Fakim, A et al.1995.
Coriandrum sativum	cotomili, corriandre.	whole plant.	Insecticidal, antifungal against Trichophytum rubrum, A. niger, F. oxysporum f.sp. Lentis.	Gurib-Fakim, A et al.1995.
Cinnamomum camphora	Ravintsara, Camphre, Camphrier, Bwa kanf, Camphor.	Leaves	Antifungal against a wide range of fungi, insecticidal, excellent remedy against broncho-pulmonary infections and flu.	Gurib-Fakim, A., Brendler, 2004.
Cinnamon verum	Canellier, Cinnamon.	Bark, leaves.	Insecticidal, antifungal, antibacterial.	Gurib-Fakim, A et l.1996.
Citrus aurantifolia	Petit limon.	Leaves, fruits.	EO extracted from fruit zest is an expectorant and antimicrobial.	Gurib-Fakim, A et al.1997.
Citrus aurantium	Bigarade, Bigaradier.	Leaves, fuits.	Antifungal against *Candida albicans*, *Aspergillus niger*, antibacterial against *E. coli*, *P. aeruginosa*, *B. subtilis*, antimycotic against *Trichoderma viridae*,	Gurib-Fakim, A et al.1997.

Table 1. (Continued)

Scientific Name	Common Name	Utilised parts	EO Properties	Reference
			Aspergillus aegypticus, *Penicillium cyclopium*.	
Citrus grandis	Pamplemoussier; Shaddock	Flowers, leaves, Fruits.	.	Gurib-Fakim, A et l. 1997.
Citrullus lantus	Melon d'eau, pastèque	seeds		Gurib-Fakim, A et l. 1996.
Citrus medica.	Citron galeux, Lime.	Leaves	Antispasmodic.	Gurib-Fakim, A et al. 1997.
Citrus reticulata	Vangassaye	Fruits,leaves.	Insecticidal.	Gurib-Fakim, A et l. 1997.
Cuminum cyminum	Cumin	Fruits.	Antifungal against *Trichophytun rubrum*, *A. niger*, *Absidia glauca*, *Alternaria alternata*, *Aspergillus niger*, *A. nidulans*, *A. fumigatus*, *Chaetomium indicum*, *Colletotrium capsici*, *Fusarium moniliformae*, *Helminthosporium oryzae*, *Phytophthora parasitica*, *Penicillium expansum*, *Pestalotia psidi*, *Rhizopus nodosus*, *Trichoderma viride*	Gurib-Fakim, A et al. 1995.
Curcuma longa	Safran vert, Safran.	Leaves, Roots.	Curcumine present in the EO obtained from the roots act on bacteria such as S. aureus, Bacillus cereus. Morphological changes have been observed due to contact with EO on Streptococci lactobacilli and Staphylococci.	Gurib-Fakim, A et al. 1997.
Cymbopogon citratus	Citronelle	whole plant.	Antibacterial against *Bacillus subtilis*, *E.coli*, *Staphylococcus aureus*.α-citral and β-citral contained in the EO of the leaves are antibacterial against both Gram positive and Gram negative	Gurib-Fakim, A et al. 1997.

Scientific Name	Common Name	Utilised parts	EO Properties	Reference
			bacteria. EO acts also on dermatophytes (*Trichophyton mentagrophytes*, *Microsporum canis*, *Epidermophyton floccosum*.	
Cynara cardunculus	Artichaut, Globe artichoke.	Leaves, roots..	Helps in intestinal disorder.	Gurib-Fakim, A et l.1995.
Cyperus rotundus	Herbe à oignon, Oumine, Nut grass	whole plant.	Diuretic, helps against diarrhoea.	Gurib-Fakim, A et l.1997.
Daucus carota	Carotte, Carotte marron	whole plant.	EO has shown an effect on the ovulation inhibition in rats. Diuretic effect observed in dogs subjects on which the EO were tested.	Gurib-Fakim, A et l.1995.
Dodonaea viscosa.	Mtsuzi mbwa, Bois de renette, Bois d'arnette, Bwa drenet.	Bark, leaves, whole plant.	Antimicrobial against *Staphylococcus aureus*, *Bacillus anthracis*, *Candida albicans*, ichyotoxic.	Gurib-Fakim, A., Brendler, 2004.
Dracaena reflexa var. angustifolia	Bois de chandelle, Sandel blan, Bois chandelle blanc, Bois chandelle rouge	Leaves	Bacteriostatic against *Staphylococcus.* Used against infectious disease.	Gurib-Fakim, A., Brendler, 2004.
Elettaria cardamomum	Elaiti, Cardamome	Seeds	EOs act on a large variety of Gram negative and Gram positive bacteria.	Gurib-Fakim, A et al.1997.
Erythrina variegata	Mourouque, Mourouc, Coral tree.	Stem, bark, leaves, flowers.	Antimicrobial, anti-pyretic, diuretic, purgative, antitumour, anthelmintic, emmenagogue.	Gurib-Fakim, A., Brendler, 2004.
Eucalyptus citriodora	Eucalyptus	Leaves	Antiseptic.	Gurib-Fakim, A et al.1996.
Eucalyptus kirtoniana	Bastard Mahogany	Leaves		Gurib-Fakim, A et al.1996.
Eucalyptus robusta	Eucalyptus rouge.	whole plant.		Gurib-Fakim, A et al.1996.
Eugenia rodriguensis	Bois fier.	Bark, leaves.	Anti-inflammatory, antibacterial.	Gurib-Fakim, A., Brendler, 2004.

Table 1. (Continued)

Scientific Name	Common Name	Utilised parts	EO Properties	Reference
Eugenia uniflora	Rousaille.	Leaves	EO acts on Dermatophytes such as *Trichophyton mentagrophytes*, *Microsporum canis*, *Epidermophyton floccosum.* EO of leaves are rich in oxygenated sesquiterpenes and has allelopathy effect.	Gurib-Fakim, A et al. 1996.
Foeniculum vulgare	Gros l'anis, Fenouil.	whole plant.	Insecticidal activity.	Gurib-Fakim, A et al. 1995.
Hedychium coronarium	Longoze, Longouze a fleurs blanches, White ginger, Gingembre de Madagascar, Fakamalao, Sakamalao, Longaza, Sakaviro, Zenzanm maron, Mon ame, Ginger lily	Roots	Insecticidal.	Gurib-Fakim, A., Brendler, 2004.
Helichrysum gymnocephalum	Immortelle à tête nue,Rambiaze, Rambiazana	Leaves	Antimicrobial against *Neisseria c.*, *Sarcina l.*, *Staphylococcus a.* Used in case of infantile diarrhoea, dysmenorrhoea, rheumatism and gout.	Gurib-Fakim, A., Brendler, 2004..
Hydnocarpus pentandra	Choulmoogra	Seeds	Oil used in treatment against skin disorders: leprosy, scabies, eczema, gout, rheumatism. Oil has wound healing properties	Gurib-Fakim, A., Brendler, 2004.
Ipomoea pes-caprae ssp. Brasiliensis	Batatran, Liane Batatran, Patatran, Poumpou, Patate à Durand, Lalanda, Vahindalalana.	Leaves, whole plant.	Antihistaminic, anti-inflammatory, Antispasmodic, neutralise toxins in poisonous fish.Treat colics, rheumatic pains, useful to alleviate muscular pains.	Gurib-Fakim, A., Brendler, 2004.
Ipomoea quamoclit	Amourette.	whole plant.	Antifungal against *Fusarium oxysporum*, Insecticidal.	Gurib-Fakim, A et al. 1995.

Scientific Name	Common Name	Utilised parts	EO Properties	Reference
Jatropha curcas	Pignon d'inde.	Seeds.	Insecticidal and purgative..	Gurib-Fakim, A., Brendler, 2004.
Juniperus bermudiana	Cyprès	Feuilles	EO is used against respiratory infections	Gurib-Fakim, A et al.1996.
Maytenus pyria	Bois à poudre.	Leaves, Stem.	Anti-inflammatory, antipyretic.	Gurib-Fakim, A et al.1995.
Melaleuca quinquenervia	Kininindrano, Oil Niaouli, Paper bark, Tea tree	Stem, leaves.	Stimulant, antispasmodic, antibacterial against Gram positive bacteria. Useful against colds, coughs and pulmonary infections.	Gurib-Fakim, A., Brendler, 2004.
Mentha x piperita	Menthe, La Menthe.	whole plant.	EO of leaves inhibit the growth and development of fungi such as *Rhizoctonia solani.*	Gurib-Fakim, A et al.1996.
Michelia champaca	Champac	whole plant.	Tonic, decrease fever.	Gurib-Fakim, A et al.1996.
Monarrhenus salicifolius.	Bois de paille-en-queue, Bois de chenilles.	Leaves	Diuretic and diaphoretic. Insecticidal	Gurib-Fakim, A., Brendler, 2004.
Murraya koenigii	Caripoulé, Carri poulet.	Leaves	EOs are present in the leaves. Used in hypertension.	Gurib-Fakim, A et al.1997.
Murraya paniculata.	Bois buis.	Leaves.	Antimicrobial effect against *Micrococcus pyogenes var. aureus*, *E. coli.*	Gurib-Fakim, A et al.1997.
Myristica fragrans	Muscadier, Muscade, Nutmeg	Fruits	EO inhibits aggregation of platelets. Insecticidal and anti-inflammatory at variable dose.	Gurib-Fakim, A et al.1996.
Ocimum basilicum L	Grand basilic, sweet basil	Leaves	Antibacterial activity *in vitro* against *B. subtilis*, *E. coli*, *P. aeruginosa*, *S. aureus and bactericidal against Salmonella typhosa.* and antifungal against *Microsporum gypseum*, *Aspergillus aeguptiacus*, *Penicilium cyclopodium*, *Trichoderma*	Gurib-Fakim, A et al.1996.

Table 1. (Continued)

Scientific Name	Common Name	Utilised parts	EO Properties	Reference
			viride,*Candida albicans*. EO also exhibited insecticidal properties against *Culex quinquefasciatus*.	
Ocimum canum	petit basilic	whole plant.	Antiseptic, expectorant, pulmonary antiseptic, inhibit fungal growth of *Rhizoctonia solani*, insecticidal.	Gurib-Fakim, A et al. 1996.
Ocimum gratissimum	Rule, Romba, Kiranjay, Basilic, Gro basilic, Basilic gran fey, Gros basilic, Basilic grande feuille.	Leaves.	Insecticidal, antibacterial, antifungal against *Microsporum canis*, *Aspergillus aegypticus*, *Penicilium cyclopium*, *Trichoderma viride*.	Gurib-Fakim, A., Brendler, 2004.
Ocimum tenuiflorum	Toolsee, Basilic à petites feuilles, Pti basilic, Sacred basilic.	Leaves.	Antibacterial, strong analgesic, anti-inflammatory and insecticidal against flies, mosquitoes and many other insects. Used on lower abdominal pains.	Gurib-Fakim, A., Brendler, 2004.
Paederia foetida	Lingue, Liane lingue, Liane caca, Laingomaimbo, Lengomena.	Leaves, whole plant.	Anti-inflammatory stronger than acetyl salicylic acid but weaker than hydrocortisone.	Gurib-Fakim, A., Brendler, 2004.
Parthenium hysterophorus	Herbe blanche.	Roots, leaves.	Febrifugal, anthelmintic.	Gurib-Fakim, A et al. 199.
Persea americana	Avocat, Avocatier	whole plant.	Phagocytic stimulant.	Gurib-Fakim, A et al. 1996.
Petroselinum crispum	Persil frisé, Persil.	whole plant.	Insecticidal. Leaf decoction helps against rheumatism.	Gurib-Fakim, A et al. 1995.
Phyla nodiflora	Verveine sauvage, Verveine	Leaves, whole plant.	Antibacterial, diuretic. Antipyretic	Gurib-Fakim, A., Brendler, 2004.
Piper betle	Betel	whole plant.	Insecticidal, Antilicrobial against *Micrococcus pyogens var. aureus*, *Bacillus subtilis*, *B. megaterium*, *Diplococcus pneumoniae*, *Streptococcus pyogenes*, *E. coli*,	Gurib-Fakim, A et l. 1997.

Scientific Name	Common Name	Utilised parts	EO Properties	Reference
			Salmonella typhosa, *Vibrio comma*, *Shigella dysentriae*, *Proteus vulgaris*, *Pseudomonas solanacaerum*, *Sarcinia lutea*, *Mycobacterium tuberuclosis*, *Erwinia carotovora* and antifungal against *Aspergillus niger*, *A. oryzae*, *Curcularia lunata and Fusarium oxysporum* and toxic against *Paramoecium caudatum*, a parasite.	
Piper borbonense	Cubebe du pays, Poivrirer marron, Bétel sauvage, Rambuu msiru, Tsiperifery.	Leaves, whole plant.	Diuretic, stimulant.	Gurib-Fakim, A., Brendler, 2004.
Pittosporum balfourii	Bois becasse, Bois begasse	Bark		Gurib-Fakim, A., Brendler, 2004.
Pittosporum senacia ssp.	Bwa zoliker, Bois jolis cœur	Leaves.	Antibcterial, anti-inflammatory, analgesic, immunoregulating.	Gurib-Fakim, A., Brendler, 2004.
Pongamia pinnata	La coqueluche, Pongam.	Seeds.	Insecticidal, antibacterial.	Gurib-Fakim, A et al.1997.
Plectranthus madagascariensis	Baume du Perou, Petite Baume, Onisme batard, Anampantsaka, Farimazo, Ranofaritra.	Leaves.	Antibacterial. Use to relieve cough and to treat fish allergies.	Gurib-Fakim, A., Brendler, 2004.
Protium obtusifolium	Bois colophane batard.	Leaves, Resin.	Anti-inflammatory.	Gurib-Fakim, A., Brendler, 2004.
Psiadia altissima.	Dingadingana	Leaves	Antimicrobial against *Staphylococcus aureus.*, *Sarcina lutea.*, *Bacillus subtilis.*, *Neisseria cinerea* , *Shigella*, *Salmonella typhi.*	Gurib-Fakim, A., Brendler, 2004.
Psiadia arguta.	Baume de l'ile Plate.	Leaves, Bark.	Anti-inflammatory, antioxydant, antiallergic, antibacterial, antiviral.	Gurib-Fakim, A., Brendler, 2004.
Psidium cattleianum	Goyave de chine jaune, goyave de chine rouge, strawberry Guava.	Leaves,fruits.	EO present in the leaves are rich in β-caryophyllene (molecule known for its ablity to modulate inflammatory processes.)	Gurib-Fakim, A et al.1996.

Table 1. (Continued)

Scientific Name	Common Name	Utilised parts	EO Properties	Reference
Psidium guajava	Mbera, Gavoala, Gavobe, Giavy, Goyavier, Goyave.	Bark, leaves, fruits.	Analgesic and anti-inflammatory. Antibacterial against Gram negative and Gram positive bacteria, antipyretic.	Gurib-Fakim, A., Brendler, 2004.
Quisqualis indica.	Badamier sauvage, Rangoon creeper, Liane vermifuge, Orientale, Santonine.	Seeds.	Anthelmintic property, antiviral properties against Sindbis virus.	Gurib-Fakim, A., Brendler, 2004.
Ravenala madagascariensis.	Ravenale, Arbre du voyageur.	Young leaves.	Antiseptic properties	Gurib-Fakim, A., Brendler, 2004.
Ravensara anisata.	Epice de Madagascar, Avozo, Havozo, Havozomamy, Havozomangidy, Havozomanitra, Laposinty, Manitranjety, Morindrano, Nois de girofle, Ravintsara, Voaravontsara.	Bark, fruit, leaves.	Shows activity against *Bacillus subtilis.*, *Bordetella bronchiseptica.*, *Escherichia coli*, *Sarcina lutea.*, *Staphylococcus aureus.*	Gurib-Fakim, A., Brendler, 2004.
Rorippa nasturtium - aquaticum	Cresson	whole plant.		Gurib-Fakim, A et al. 1995.
Rosmarinus officinalis L.	Romarin.	whole plant.	EO has an effect on cardiac muscles by relaxing them. Leaves infusion used to keep hair healthy.	Gurib-Fakim, A et al. 1996.
Schinus terebinthifolius	Baie rose, faux poivrier, Brazilian pepper tree.	Leaves, Bark.	Antimicrobial against Gram positive bacteria and Mycobacteria	Gurib-Fakim, A., Brendler, 2004.
Senecio bojerii	Trezi	Roots	Antioxydative: superoxide and hydroxy radical-scavenging activities	Gurib-Fakim, A., Brendler, 2004.
Siphonochilus aethiopicus	African ginger root, African ginger, gingembre Africaine, zenzero Africano.	Roots	Antibacterial against Gram positive bacteria, anti-inflammatory, uterine relaxing.	Brendler et al. 2010.

Scientific Name	Common Name	Utilised parts	EO Properties	Reference
Syzygium aromaticum	Giroflier, Clove tree.	Leaves, fruits.	Antibacterial, antifungal, insecticidal, Carminative, antiseptic, analgesic, use in gingivitis. EO has an effect on *Anopheles stephensi*. EO is also very effective in the inhibition of platelet aggregation.	Gurib-Fakim, A et al. 1996.
Tagetes erecta L.	Pissenlit, Guinda, African marigold, Œillet.	whole plant.	EO: Antifungal, anti-allergenic.	Gurib-Fakim, A et al. 1995.
Tambourissa leptophyla	Troundra, Djarou, Mbotsa.	Fruit zest.	Antifungal against *Cladosporium sp.*	Gurib-Fakim, A., Brendler, 2004.
Tanacetum parthenium.	Camomille.	whole plant.	Use in relation to insomnia, anxiety, headache, nervous depression.	Gurib-Fakim, A et al. 1995.
Terminalia bentzoe	Bois charon, badamier à petites feuilles.	Leaves, Bark.	Bacteriostatic and activity against dysentry.	Gurib-Fakim, A., Brendler, 2004.
Tetradenia riparia.	Ginger bush, Baume, Bombe, Petite Baume, Beaume.	Leaves.	Antibacterial, antispasmodic expectorant, purgative. Use to treat sinusitis, colds, eczema.	Gurib-Fakim, A., Brendler, 2004.
Thespesia populnea	Porcher, Bwadroz, Portia tree, Bois de rose, Mahoe,Valo, Valorao,Varo, Valirano.	Bark, Leaves, Sap, whole plant.	Antiamoebic, Antibacterial activity against both Gram negative and Bram positive bacteria.	Gurib-Fakim, A., Brendler, 2004.
Toddalia asiatica	Patte poule à piquants, Bambara, Ronce, Patte poule, Petit Patte poule, Patte poule piquant, Liane patte de poulet.	Leaves.	Antibacterial activity against *E. coli*, *K. pneumoniae*, *P. aeruginosa*, *S. newport*, *S. aureus* and antifungal activity against *A. fumugatus*, *A. niger*, *Microspermum gypsum*, *Rhizophus stolonifer*. Used for fever, coughs and influenza, as a stimulant, as a stomach remedy, against rheumatism and syphilis	Gurib-Fakim, A., Brendler, 2004.
Tropaeolum majus	Capucine, Nasturium	Leaves	Antibacterial properties on a variety of Gram negative and Gram positive bacteria. Fongistatic.	Gurib-Fakim, A et al. 1997.
Triphasia trifolia	Orangine	Leaves.	EO obtained from leaves. Leaves decoction used for flu and palpitations.	Gurib-Fakim, A et al. 1997.

Table 1. (Continued)

Scientific Name	Common Name	Utilised parts	EO Properties	Reference
Veteria seychellarum.	Bwad fer, Bois de fier.	Leaves	Use to treat acute bronchitis and in rheumatic ailments.	Gurib-Fakim, A., Brendler, 2004.
Vetiver zizanioides.	Vetiver, Kus-kus grass.	whole plant.	Antifungal, insecticidal.	Gurib-Fakim, A et al.1997.
Vitex trifolia	Bois cabri, Lilas de Perse, Samaroo, Kalimirefatra, Bois noux noux.	Leaves, Stem.	EO isolated from leaves, and has antibacterial properties against *B. subtilis*, *S. albus*, *P. aeruginosa*, *K. pneumoniae.* Use for headache, hypertension, rheumatisimal pains and sprains.	Gurib-Fakim, A., Brendler, 2004.
Xanthium strumarium	Grosse pagode, bogues..?	Leaves, Stem, Roots.	Diuretic, sedative, use to treat herpes, and urinary infection.	Gurib-Fakim, A et al.1995.
Zingiber officinale	Gingembre	Roots	Hypoglycemic, fongicide, acts *in vitro* on Gram positive and Gram negative bacteria. Gingerol and shogoal molecules present in the roots has antipyretic and analgesic effect	Gurib-Fakim, A et al.1997.

Table 2. Overview of major components present in EOs, their biological properties and applications

Properties	Molecules	Major composition of EO	Classification	Other Uses	Ref.
Antioxidant	Linalool	*Ocimum basilicum* (Sweet basil) *Cinnamomum altissimum (Mountain cinnamon)*	Terpene alcohol	Fixative agent, floral and non-floral fragrance, use in the synthesis of Vitamin E.	Hussain et al. 2008.
	Eugenol	*Syzygium aromaticum* (Clove) *Ocimum basilicum* (Sweet Basil)	Phenol	In fragrance and flavors, as anesthetic for tropical fish, in dentistry.	Porres-Martínez et al. 2013
	Eucalyptol	*Melaleuca armillaris Piper betel* (Betel*) Melaleuca alternifolia* (Tea tree)	Alcohol	To make medicated balm, fragrances, dental preparations.	Chabir et al. 2011
	Thymol	*Thymus vulgaris* (Thyme) *Salvia lavandulifolia* (Sage)	Phenol	Fungicide; fragrance in soap ad household products.	Shan et al. 2005; Wojdylo et al. 2007
	Rosmanol	*Rosmarinus officinalis* (Romarin)	Diterpenes	-	Wang 2003
Anti Bacterial	Carvacrol and Thymol	*Oreganum vulgare* (Oregano) *Thymus vulgaris* (Thyme)	Carvacrol: Phenol Thymol: Phenol	Carvacrol : in dental preparations, Thymol : In cosmetics, in traces for herbaceous fragrances, as fungicide	Burt 2004; Ultee et al. 2002; Lambert et al. 2001
	Eugenol	*Syzygium aromaticum* (Clove) *Laurus nobilis* (Bay leaves) *Cananga odorata*(ylang ylang) *Pogostemon cablin*(patchouli) *Jasminum grandiflorum* (jasmine) *Myristica fragrans* (Nutmeg).	Phenol	In fragrance and flavors, as anesthetic for tropical fish, in dentistry.	Maree et al. 2014; Burt, 2004; Wendakoo and Sakaguchi, 1995.
	p-Cymene and Carvacrol	*Ocimum basilicum(*Sweet basil) *Thymus vulgaris* (Thyme)	Carvacrol: Phenol *p*-Cymene: alkylbenzene	*p*-Cymene : analgesic	Solórzano-Santos and Miranda-Novales 2012; Veldhuizen 2006; **Kiskó and Roller 2005**
	Carvone	*Enterolobium contortisiliquum* (Pacara earpode) *Mentha spicata* L (Spearmint)	Ketone	Food flavors and fragrances	Shahat et al. 2008; Aggarwal et al. 2002.

Table 2. (Continued)

Properties	Molecules	Major composition of EO	Classification	Other Uses	Ref.
	Estragole	*Enterolobium contortisiliquum* (Pacara earpode) *Artemisia dracunculus* (Tarragon) *Ocimum basilicum* (sweet basil) *Illicium verum* (Star anise)	Aromatic ether	Perfumes and Food additives	Ariamuthu et al. 2013
	Cinnamaldehyde	*Cinnamomum verrum* (cinnamon)	Aldehyde	Agrichemical (fungicide), food flavors, Perfumes.	Yossa et al. 2012.
	Terpinene	*Melaleuca alternifolia* (Narrow leaved tea-tree) *Cuminum cyminum* (Cummin)	Monoterpene hydrocarbon	Fragrance	Oyedemi et al. 2009; Dorman and Deans 2000.
Anticancer	β-elemene	*Boswellia* (Frankincense)	Sesqui-terpenes	Insect pheromones	Chen et al. 2013
Antiacetyl-cholinesterase	Isoeugen, Eugenol	Eucalyptus sp.	Phenyl-propene	Fragrance	Yeom et al. 2013

CONCLUSION

Currently, there is a renewed interest in the use of EO and plant based extracts. It is thus becoming more and more important to have accurate and reliable information on EOs composition which have been revered for their chemical diversity. Also, the foundation of organoleptic properties and biological properties of EOs are tracked by their respective compositions. Moreover, post-harvest factors among many others can result in the alteration of plant metabolites, which can lead to drastic change in molecule structures. Hence, without regular steadfast testing of the composition of EOs, there is no guarantee of the product being used and its impact on the health.

Microwave and ultrasonic-assisted extraction seemed to have gained success and some literature reviews referred to them as established methods of extraction with maximum efficacy and minimum extraction time. Also, the new trend is focused on green technologies, which are less detrimental to the environment, and can at the same time increase the quality of the end products. These green technologies are nowadays prioritized when it comes to extraction requirements.

REFERENCES

Abert Vian, M., Fernandez, X., Visinoni, F. and Chemat, F. (2008). "Microwave hydrodiffusion and gravity, a new technique for extraction of essential oils". *Journal of Chromatography A*, *1190*, 14-17.

Amirah, D. M., Reddy P. and Maksudur, R. khan. (2012). "Comparison of extraction Techniques on extraction of gallic acid from stem bark of Jatropha curcas". *Journal of Applied Sciences*, *12*, 1106- 1111.

Bagheri, H., Abdul Manap, M. Y. B. and Solati, Z. (2014). "Antioxidant activity of Piper nigrum L. essential oil extracted by supercritical CO_2 extraction and hydro-distillation". *Talanta*, *121*, 220-228.

Bakkali, F., Averbeck, S., Averbeck, D. and Idaomar, M. (2008). "Biological effects of Essential oils-A review". *Food and Chemical Toxicology*, *46*, 446-475.

Bansal, A., Chhabra, V., Rawal, R. K. and Sharma, S. (2014). "Chemometrics: A new scenario in herbal drug standardization." *Journal of Pharmaceutical Analysis*, *4*, 223-233.

Beckman. E. J. (2004). "Supercritical and near-critical CO_2 in green chemical synthesis and processing". *Journal of Supercritical Fluids*, *28*, 121-191.

Bi, W; Zhou, J. and Row, KH. (2011). "Decaffeination of coffee bean waste by solid-liquid extraction". *Korean Journal of Chemical Engineering*, *28*, 221–224.

Binello, A., Orio, L., Pignata, G., Nicola, S., Chemat, F. and Cravotto. G. (2014). "Effect of microwaves on the *in situ* hydrodistillation of four different Lamiaceae". *Comptes Rendus Chimie.*, *17*(3), 181–186.

Brendler, T., Eloff, J. N., Gurib-Fakim, A. and Phillips, L. D. (2010). "African Herbal Pharmacopoeia" Association for African Medicinal Plants Standards (AAMPS), 1st ed. Viva Publication New Delhi: India.

Chaiyana, W. and Okonogi, S. 2012. "Inhibition of cholinesterase by essential oil from food plant". *Phytomedicine*, *19*, 836-839.

Chen, Q., Hu, X., Wang, Y., Gan, Z., Jiang, S., Liu, P., Dai, Y. and Ni, Y. (2012). "Enrichment of cuminaldehyde and p-mentha-1,4-dien-7-al in cumin (Cuminum cyminum L.) oil by molecular distillation". *Separation and Purification Technology, 98*, 136-144.

Da Porto, C., Decorti, D. and Kikic, I. (2009). "Flavour compounds of Lavandula angustifolia L. to use in food manufacturing: comparison of three different extraction methods". *Food Chemistry., 112*, 1072-1078.

Delamare, A. P. L., Moschen-Pistorello, I. T., Artico, L., Atti-Serafini, L. and Echeverrigaray, S. (2007). "Antibacterial activity of the essential oils of Salvia officinalis L. and Salvia triloba L. cultivated in South Brazil". *Food Chemistry., 100*, 603-608.

Desty, D. H., Goldup, A. and Swanton, W. T. (1962). In *Gas Chromatography*, edited by Brenner, N., Callen, J.E., and M.D. Weis, 102-105, London: Academic Press.

Dixon, D. J. and Johnston, K. P. (1997). *"Supercritical Fluids"*. In *Encyclopedia of Separation Technology*, edited by Ruthven, D. M., 1544-1569, New Jersey: John Wiley.

Ferhat, M. A., Brahim, Y. and Chemat, F. (2007). "Comparison of different isolation methods of essential oil from Citrus fruits: cold pressing, hydrodistillation and microwave 'dry' distillation". *Flavour and Fragrance Journal, 22*(6), 494-504.

Ferhat, M. A., Meklati, B. Y., Visinoni, F., Abert Vian, M. and Chemat, F. 2008. "Solvent free microwave extraction of essential oils Green chemistry in the teaching laboratory". *Chimica Oggi, 26*, 48–50.

Ferhat, A., Fabiano-Tixier, A. S., El Maataoui, M., Maingonnat, J. F., Romdhane, M. and Chemat, F. (2011). "Microwave Steam Diffusion for extraction of essential oil from orange peel: Kinetic data, extract's global yield and mechanism". *Food Chemistry., 125*(1), 255-261.

Golmakani, M. T. and Rezaei, K. (2008). "Comparison of microwave-assisted hydrodistillation with the traditional hydrodistillation method in the extraction of essential oils from Thymus vulgaris L". *Food Chemistry, 109*, 925-930.

Guenther, E. (1948). *The Essential Oils Volume One: History-Origin in Plants-Production - Analysis*. New York: D. Van Nostr and Company, Inc.

Gurib-Fakim, A., Guého, J. and Bissondoyal, M. D. (1995). *"Plantes médicinales de Maurice"*. Tome 1 Edition de l'océan Indien, Rose Hill (Maurice).

Gurib-Fakim, A., Gueho, J. and Sewraj, M. D. (1996). *"Plantes médicinales de Maurice"*. Tome 2. Editions de L'ocean Indien, Stanley, Rose Hill, Mauritius.

Gurib-Fakim, A., Guého, J. and Bissondoyal, M. D. (1997). *"Plantes médicinales de Maurice"*. Tome 3. Edition de l'Océan indien, Rose Hill, Mauritius.

Gurib-Fakim, A. and Brendler, T. (2004). *"Medicinal and aromatic plants of the Indian Ocean Islands"*. Medpharm Scientific Publishers, Stuttgart, Germany.

Hartono, R., Mansoori, G. A. and Suwono, A. (2001). "Prediction of solubility of biomolecules in supercritical solvents". *Chemical Engineering Science, 56*, 6949-6958.

Hunter, V. M. (2009). "Essential Oils: Art, Agriculture, Science, Industry and Entrepreneurship a focus on the Asia-Pacific region". Malaysia: University of Malaysia Perli.

Jayanthi, P. and Lalitha, P. (2013). "Comparison of conventional and sound assisted methods for extraction of Eichhornia crassipes (Mart.) solms". *Asian Journal of Pharmaceutical and Clinical Research, 6*, 143-146.

Lindeman, Z., Waggoner, M., Batdorff, A. and Humphreys, T. (2014). "Assessing the antibiotic potential of essential oils against Haemophilus ducreyi". *BMC Complementary and Alternative Medicine*, *14*, 172-178.

Marriott, P. J., Shelliea, R. and Cornwell, C. (2001). "Gas chromatographic technologies for the analysis of essential oils". *Journal of Chromatography A*, *936*, 1–22.

Martins, P. F., Carmona, C., Martinez, E. L., Sbaite, P., Filho, R. M. and Maciel, M. R. W. (2012). "Short path evaporation for methyl chavicol enrichment from basil essential oil" *Separation and Purification Technology*, *87*, 71–78.

Mondello, L., Casilli, A., Quinto Tranchida, P., Costa, R., Dugo, P. and Dugo, G. (2004). "Fast GC for the Analysis of Citrus Oils". *Journal of Chromatographic Science*, *42*, 410-416.

Mussatto, S. I., Ballesteros, L. F., Martins, S. and Teixeira, J. A. (2011). "Extraction of antioxidant phenolic compounds from spent coffee grounds". *Separation and Purification Technology*, *83*, 173-179.

Navarrete, A., Wallraf, S., Mato, R. B. and Cocero, M. J. (2011). "Improvement of Essential Oil Steam Distillation by Microwave Pretreatment". *Industrial and Engineering Chemistry Research*, *50*(8), 4667-4671.

Newman, D. J. and Cragg, G. M. (2007). "Natural products as sources of new drugs over the last 25 years." *Journal of Natural Products*, *70*(3), 461-477.

Okoh, O. O., Sadimenko, A. P. and Afolayan, A. J. (2010). "Comparative evaluation of antibacterial activities of essential oils of Rosmarinus officinalis L. obtained by hydrodistillation and solvent free microwave extraction methods". *Food Chemistry*, *120*(1), 308-312.

Olmedo, R., Nepote, V. and Grosso Nelson, R. (2014). "Antioxidant Activity of Fractions from Oregano Essential Oils Obtained by Molecular Distillation" *Food Chemistry*, *156*, 212-219.

Orio, L., Cravotto, G., Binello, A., Pignata, G., Nicolab, S. and Chemat, F. (2012). "Hydrodistillation and in situ microwave generated hydrodistillation of fresh and dried mint leaves: a comparison study". *Journal of the Science of Food and Agriculture*, *92*, 3085-3090.

Ozek, T. and Demirci, F. (2012). "Isolation of natural products by preparative gas chromatography". *Methods in Molecular Biology*. *864*, 275-300.

Pakniyat, E. and Mousavi, M. (2014). "Improvement of GC-MS analysis of Shahrbabak Ziziphora tenuior Essential oil by using the Multivariate Curve Resolution Approaches". *Journal of the Chinese Chemical Society*, *61*, 649–658.

Park, E. H., Yang, S. O., Hyun, S. H., Park, J. S., Choi, H. K. and Marriott, J. P. (2012). "Simple preparative gas chromatographic method for isolation of menthol and menthone from peppermint oil, with quantitative GC-MS and 1H NMR assay" *Journal of Separation Science*, *35*, 416–423.

Périno-Issartier, S., Ginies, C., Cravotto, G. and Chemat, F. (2013). "A comparison of essential oils obtained from lavandin via different extraction processes: Ultrasound, microwave, turbohydrodistillation, steam and hydrodistillation". *Journal of Chromatography A*, *1305*, 41-47.

Raina, R. and Hall, P. (2008). "Comparison of Gas Chromatography-Mass Spectrometry and Gas Chromatography-Tandem Mass Spectrometry with Electron Ionization and

Negative-Ion Chemical Ionization for Analyses of Pesticides at Trace Levels in Atmospheric Samples". *Analytical Chemistry Insights*, *3*,111–125.

Rubiolo, P., Liberto, E., Sgorbini, B., Russo, R., Veuthey, J. L. and Bicchi, C. (2008). "Fast-GC–conventional quadrupole mass spectrometry in essential oil analysis". *Journal of Separation Science*, *31*, 1074-1084.

Sahraoui, N., Abert Vian, M., Bornard, I., Boutekedjiret, C. and Chemat, F. (2008). "Improved microwave team distillation apparatus for isolation of essential oils Comparison with conventional steam distillation". *Journal of Chromatography A*, *1210*, 229-233.

Suetsugu, T., Tanaka, M., Iwai, H., Matsubara, T., Kawamoto, Y., Saito, C., Sasaki, Y., Hoshino, M., Quitain, A. T., Sasaki, M., Sakamoto, J. and Goto, M. (2013). *"Supercritical CO_2 extraction of essential oil from Kabosu (Citrus sphaerocarpa Tanaka) peel"*. *Flavour*, *2*, 18.

Tranchida, P. Q., Zoccali, M., Franchina, F. A., Bonaccorsi, I., Dugo, P. and Mondello, L. (2013). "Fast gas chromatography combined with a high-speed triple quadrupole mass spectrometer for the analysis of unknown and target citrus essential oil volatiles". *Journal of Separation Science*, *36*(3), 511-516.

Tovar, L. P., Maciel, M. R. W., Pinto, G. M. F., Filho, R. M. and Gomes, D. R. (2010). "Factorial design applied to concentrate bioactive component of Cymbopogon citratus essential oil using short path distillation". *Chemical Engineering Research and Design*, *88*, 239-244.

Uysal, B., Sozmen, F. and Buyuktas, B. S. (2010). "Solvent-free Microwave Extraction of Essential Oils from Laurus nobilis and Melissa officinalis: Comparison with Conventional Hydrodistillation and Ultrasound Extraction". *Natural Product Communications*, *5*, 111-114.

Wang, W., Wu, X., Han, Y., Zhang, Y., Sun, T. and Dong, F. (2011). "Investigation on ultrasound-assisted extraction of three dibenzylbutyrolactone lignans from Hemistepta lyrata". *Journal of Applied Pharmaceutical Science.*, *1*(9), 24-28.

Xiao, J., Chen, J. and Xu, M. (2007). "Supercritical fluid CO_2 extraction of essential oil from Marchantia convoluta: global yields and extract chemical composition". *Electronic Journal of Biotechnology*, *10*(1), 1-7.

Xiong, Y., Zhao, Z., Zhu, L., Chen, Y., Ji, H. and Yang, D. (2013). "Removal of three kinds of phthalates from sweet orange oil by molecular distillation, LWT". *Food Science and Technology*, *53*(2), 487-491.

Yang, F. Q., Wang, H. K., Chen, H., Chen, J. D. and Xia, Z. N. (2011). "Fractionation of volatile constituents from Curcuma rhizome by preparative gas chromatogaphy". *Journal of Automated Methods and Management in Chemistry*, 2011, 1-6.

Yang, Y. C., Wei, M. C. and Hong, S. J. (2014). "Ultrasound-assisted extraction and quantitation of oils from Syzygium aromaticum flower bud (clove) with supercritical carbon dioxide". *Journal of Chromatography A*, *1323*, 18–27.

Zu, G., Zhang, R., Yang, L., Ma, C., Zu, Y., Wang, W. and Zhao, C. (2012). "Ultrasound-Assisted Extraction of Carnosic Acid and Rosmarinic Acid Using Ionic Liquid Solution from Rosmarinus officinalis". *International Journal of Molecular Sciences, 13*(9), 11027-11043.

Zuo, H. L., Yang, F. Q., Zhang, X. M. and Xia, Z. N. (2012). "Separation of cis- and trans-asarone from Acorus tatarinowii by preparative gas chromatography". *Journal of Analytical Methods in Chemistry*, 2012, 1-5.

In: Essential Oils
Editor: Miranda Peters

ISBN: 978-1-63484-351-5

Chapter 4

EVALUATION OF ESSENTIAL OIL OF *OCIMUM GRATISSIMUM* L. FOR ANTIBACTERIAL ACTIVITY AND POTENTIAL TO MODIFY AMINOGLYCOSIDE TOXICITY

José J. S. Aguiar[1], Cicera P. B. Sousa[1], Fernando G. Figueredo[1], Vanessa C. N. Bitu[1], Francisco A. V. Santos[1], Maria I. Ferreira[1], Maria A. da Silva[1], Flórido S. N. Peixoto[1], Henrique D. M. Coutinho[1,2,*] and Edinardo F. F. Matias[1]

[1]Faculdade Leão Sampaio, Juazeiro do Norte (CE), Brasil
[2]Departamento de Química Biológica, Universidade Regional do Cariri, URCA, Crato-CE, Brasil

ABSTRACT

Bacteria are capable of causing severe infections, adapting to divergent situations and developing resistance to antibiotics. Products extracted from medicinal plants can be utilized as antibacterial agents and drug-resistance modifiers, thereby potentiating or mitigating the activity of synthetic antibiotics. The family Lamiaceae is characterized as having species with pharmacological potential, particularly with respect to combating bacterial infections. The aim of the present work was to evaluate the antibiotic and antibiotic-modifying activity of essential oil of *Ocimum gratissimum* Lamiaceae, using the microdilution technique. Standard and multiresistant bacterial strains of *Staphylococcus aureus, Escherichia coli and Pseudomonas aeruginosa* were utilized, and antibiotics of the aminoglycoside class were tested. With respect to evaluating the antibacterial effect of the *O. gratissimum* L, they were found to have a minimum inhibitory concentration (MIC) $\geq$ 1024 μg/mL, except *S. aureus* ATCC 25923 with

* E-mail: hdmcoutinho@gmail.com, Phone: +55(88)31021212, Fax: +55(88) 31021291.

MIC = 512 µg/mL. The drug-modifying potential of the essential oil of *O. gratissimum* L. was found to be antagonistic, reducing the effect of antibiotics, against the bacterial strains tested, except against *E. coli* when combined with amikacin and against *P. aeruginosa* combined with gentamicin, where there was no effect. The data obtained are promising, but further studies are needed to isolate the active compounds and to conduct pharmacological tests *in vivo*, making it possible to develop new therapeutic alternatives for the treatment of diseases caused by multiresistant microorganisms.

Keywords: *Ocimum gratissimum,* antibiotic-modifying activity, essential oil

Introduction

Bacteria are prokaryotic microorganisms that reproduce by binary fission, and they have filamentous double-strand circular DNA and are differentiated by their cell wall, where they are classified as Gram-positive or Gram-negative. (Murray et al., 2004). These microscopic beings are usually found on the surface of the skin, mucosae and intestinal tract of humans and animals. The genus *Staphylococcus* is distributed in nature and occurs in the normal microbiota (Coutinho et al., 2009), where the bacteria are usually found as single cells, doublets, or chains and are determinately classified as an important human pathogen, causing diseases with a fatal prognosis (Vermelho et al., 2007).

The bacterial species *Escherichia coli* is classified as a microorganism capable of causing severe infections and is associated with a variety of diseases, included sepsis, urinary tract infection (UTI), meningitis and gastroenteritis, while *Pseudomonas aeruginosa,* is a Gram-negative bacillus with polar flagella, which confer motility, where it cause various cutaneous infections, such as those from severe burns, and endocarditis in immunocompromised patients (Murray et al., 2004).

The ability to develop resistance to antimicrobial agents is a characteristic observed among microorganisms in general. However, bacteria are able to develop different mechanisms of resistance that are genetically coded. Resistance genes can be acquired through DNA mutation and transfer (Carneiro, 2006).

With the increase in incidence of resistance to antibiotics, substances derived from plants, can be an interesting alternative (Oliveira et al., 2007a; Silva et al., 2007). Considering that natural products of plant origin can alter the action of antimicrobial agents, either by enhancing or reducing the activity of the drug (Coutinho et al., 2008), in the last years, many plants have been evaluated not only for their antibacterial effects but also antibiotic-modifying activity (Simon, 2004). Medicinal plants are therapeutic resources used by traditional cultures for various illnesses and conditions. They have pharmacological actions that have aroused the interest of scientists, for the purpose of developing studies directed at traditional and popular medicine (Alvim et al., 2006).

Accordingly, the family Lamiaceae has a variety of species commonly utilized in the cosmetic and pharmaceutical industry (Sousa and Lorenzi, 2005). One of these species is *Ocimum gratissimum* Lamiaceae, a hardy perennial plant, easily found in all of Brazil and popularly known as alfavaca (Matos, 1998), It is utilized in the treatment of illnesses that affect the airways and others, and it has a chemical composition high in eugenol (Pereira et al., 2004). Pharmacological assays have shown antioxidant, anticonvulsant, anesthetic, dental analgesic, fungicidal (Sartoratto et al., 2004), and antimicrobial activities (Matias et al., 2010).

The aim of the present study was to evaluate the antibiotic and antibiotic-modifying activity of essential oils of *Ocimum gratissimum* Lamiaceae and *Plectranthus amboinicus* Lamiaceae, alone and combined against standard and multiresistant bacterial strains.

METHODOLOGY

As strains bacterial utilized in the present study were strains of *S.aureus* (SA-ATCC 25923 and SA358), *E.coli* (EC- ATCC 10536 and EC 27) and *P. aeruginosa* (PA-ATCC 15442 and PA03) with the resistance profile identified in Table 1.

All strains were maintained on slants with heart infusion agar (HIA, Difco Laboratories LTDA). Before the assay, the cells were grown for 24 h at 37°C in brain heart infusion broth (BHI, Difco Laboratories LTDA).

The leaves of *Ocimum gratissimum* L.were collected according to the ethnopharmacological place, around 08:00 h, in the garden of medicinal plants of Universidade Regional do Cariri – URCA, located in the municipality of Crato - Ceará, Brazil, geographic coordinates of 7° 13' 46" S, 39° 24' 32" W. A dried specimen was deposited in the Herbario Caririense Dardano de Andrade Lima, of Universidade Regional do Cariri – URCA, under Nos. 3978 and 26433, respectively.

Table 1. Origin of bacterial strains and antibiotic resistance profile

Bacterial strain	Origen	Resistance profile
Staphylococcus aureus SA 358	Surgical wound	Oxa, Gen, Tob, Ami, Can, Neo, Para, But, Sis, Net
Staphylococcus aureus ATCC 25923	–	–
Escherichia coli EC27	Surgical wound	Ast, Ax, Amp, Ami, Amox, Ca, Cfc, Cf, Caz, Cip, Clo, Im, Can, Szt, Tet, Tob
Escherichia coli ATCC 10536	–	–
Pseudomonas aeruginosa PA03	Catheter tip	Cpm, Ctz, Im, Cip, Ptz, Lev, Mer, Ami
Pseudomonas aeruginosa ATCC 15442	–	–

Ast-Aztreonan; Ax- Amoxacillin; Amp-Ampicillin; Ami-Amikacin; Amox-Amoxillin, Ca-Cefadroxil; Cfc-Cefaclor; Cf- Cephalothin; Caz-Ceftazidime; Cip-Ciproflaxacin; Clo –Chloramphenicol; Im-Imipenem; Can-Kanamycin; Szt-Sulfamethotrim, Tet-Tetracycline; Tob- Tobramycin; Oxa-Oxacillin; Gen-Gentamicin; Neo- Neomycin; Para- Paramomycin; But- Butirosin; Sis-Sisomicin; Net- Netilmicin; (-) absence of resistance or non-significant resistance.

The leaves of *O. gratissimum* L were triturated, made into 400-g amounts each, placed separately in glass flasks, submersed in distilled water and boiled for 2 h in a Clevenger type oil extractor (Silva, 2011). The water/oil mixture obtained of was separated, treated with anhydrous sodium sulfate and filtered. The oils were prepared at a concentration of 10 mg/mL, dissolved in DMSO (dimethyl sulfoxide), and then diluted with distilled water to a concentration of 1024 μg/mL (Coutinho et al., 2008).

The identification of the classes of secondary metabolites in the essential oil of *O, gratissimum* L. was done using a Shimadzu QP2010 series GC/MS system equipped with an Rtx-5MS capillary column (30 m x 0.25 mm i.d., 0.25 μm film thickness); helium was the carrier gas at 1.5 mL/min, flow rate was 0.8 mL/min, using split mode. The injector of injector and detector temperatures were 250 and 200°C, respectively. The column temperature was programmed from 35 to 180°C at 4°C/min and then 180 to 250°C at 10°C/min. The mass spectrometer was operated at an ionization energy of 70 eV. The identification of individual components was based on their fragmentations in the NIST 8 mass spectral library and through comparison with literature data (Adams, 2001).

MIC (minimum inhibitory concentration) was determined using a microdilution assay in broth medium (NCCLS, 2003), with an inoculum of 100 μL of each strain in 96-well microtiter plates. The cell suspension was in BHI broth and had a starting concentration of 10^5 CFU/mL, with 2-fold serial dilutions. Solutions of samples and of their respective combinations (100 μL) were added to each well. The final concentrations varied 512 - 8 μg/mL. The controls were with the standard antibiotics amikacin and gentamicin, whose final concentrations varied 2500– 2.4 μg/mL. The plates were incubated at 35°C for 24 h. Afterwards, viable cells were stained with resazurin and absorbance was read (Salvat et al., 2001). The MICs were recorded as the lowest concentration needed to inhibit growth.

Antibiotic-modifying activity was determined by the method described by Coutinho et al. (2008), where the as solutions of essential oils and their combinations were tested at sub-inhibitory concentrations (MIC/8). A volume of 100 μL of the solution containing 10% BHI, inoculum and sample was added to each will in the alphabetic direction of the plate. Next, 100 μL of drug were added to the first well, followed by 2-fold serial microdilution up to the next to last well.

The concentrations of aminoglycosides varied gradually from 5000 to 2.44 μg/mL. The plates were incubated at 35°C for 24 h, and afterwards, viable cells were determined by resazurin staining.

The results of tests were obtained in triplicate and expressed as geometric mean. Statistical analysis was carried out using ANOVA followed by Bonferroni post-tests, utilizing GraphPad Prism 6.0 (Matias, 2013), and $p < 0.05$ was considered significant.

RESULTS AND DISCUSSION

The yield of essential oil from *O. gratissimum* L. was 0.45%, compatible with the finding of Vieira et al. (2001), and chromatographic analysis showed that the major constituent was eugenol (65%) (Table 2), in agreement with the study by Sartoratto et al. (2004). The Figure 1 shows the results of evaluating the modulatory activity of oils when combined with amikacin and gentamicin.

Table 2. Chemical composition (%) of essential oil of *Ocimum gratissimum* L.

Componentes	**Tr(min)**[a]	(%)
β-pinene	3.58	1,59
β-ocimeno	4.09	4,47
p-cineol	4,15	15,17
β-linalol	4.85	0,37
α-terpineol	6,15	0,73
Eugenol	8,42	65,26
β-elemeno	8,90	0,66
(E)-cariofileno	9,37	3,04
α-humuleno	9,82	0,47
germacreno D	10,17	1,91
β-selineno	10,26	4,40
γ-gurjuneno	10,37	1,57
Total		99,64

[a] Retention time.

Table 3. Minimum inhibitory concentration (MIC) of essential oil of *Ocimum gratissimum* (μg/mL)

Essential oil	EC27	EC-ATCC 10536	SA358	SA-ATCC 25923	PA03	PA-ATCC 15442
EOOG	≥1024	≥1024	≥1024	512	≥1024	≥1024

EOOG – Essential oil of *Ocimum gratissimum*; EOOG. EC – *Escherichia coli*, SA – *Staphylococcus aureus*, PA- *Pseudomonas aeruginosas*.

The results demonstrated that the combinations of EOOG with aminoglycosides demonstrated antagonism, with exception of the combination with gentamicin against PA03 and with amikacin against EC 27, but without significance.

The evident antagonistic action observed with EOOG can be explained by the chelating property of secondary metabolites such as eugenol in the essential oil in study, capable of causing mutual chelation. This effect possibly accounts for the reduction in the activity of antibiotics in the presence of EOOG (Behling et al., 2004).

However, these findings are at odds with the results of Escobar (2002) who described the major constituent eugenol as aphenylpropanoid compound with substantial antibacterial activity.

On the basis of the results obtained, we conclude that However, the combination of antibiotics with EOOG resulted in an antagonistic effect, decreasing the antimicrobial activity of aminoglycosides. Further studies are needed to isolate the active compounds and develop new therapeutic alternatives for the treatment of diseases caused by multiresistant microorganisms.

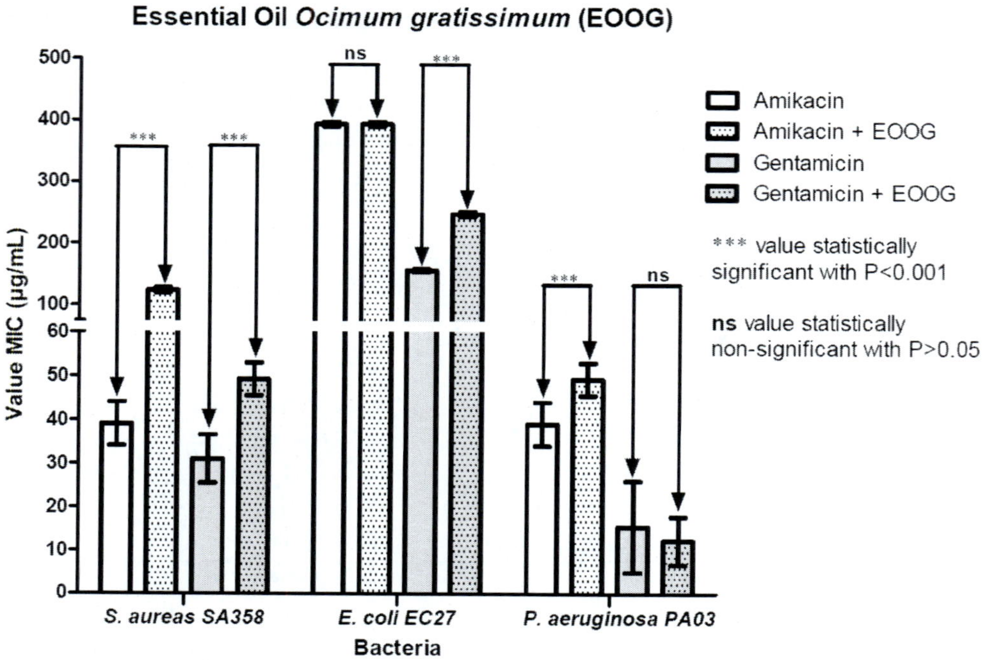

Figure 1. Comparative analysis of the association between the essential oil of *Ocimum gratissimum* (EOOG) with amikacin and gentamicin.

Acknowledgments

The authors are thankful for the support and cooperation from the FaculdadesLeãoSampaio FALS-CE, URCA/ LPPN/LMBM/LFQM Universidade Regional do Cariri-CE/Laboratory of Regional Natural Products Research /Laboratory of Microbiology and Molecular Biology. Dr. A. Leyva helped with English translation and editing of the manuscript.

References

Adams, R. P. (2001). Identification of essential oil components by gas chromatography/quadrupole mass spectroscopy. Carol Stream, Allured Publishing Corporation, USA.

Alvim, N. A. T. I., Ferreira, M. A., Cabral, I. E., Filho, A. J. A. (2006). The use of medicinal plants as a therapeutical resource: from the influences of the professional formation to the ethical and legal implications of its applicability as an extension of nursing care practice. *Rev. Latino-americana de Enfermagem.*, 14, 3.

Behling, E. B., Sendão, M. C., Franscescato, H. D. C., Antunes, L. M. G., Bianchi, M. L. P. (2004). Flavonoid quercetin: general aspects and biological actions. *Alimentary Nutrition*, 15, 285-292.

Carneiro, J. C. O. (2006). Antibacterial use pattern in an CCU: correlatiojns with the bacterial drug resistance. MSC Thesis. Brasília, Universidade de Brasília, Brasil.

Coutinho, H. D. M., Costa, J. G. M., Lima, E. O., Falcão, S. V. S., Siqueira, J. J. P. (2009). Herbal therapy associated with antibiotic therapy: potentiation of the antibiotic activity against methicillin – resistant Staphylococcus aureus by Turnera ulmifolia L. BMC Complem. *Altern. Med.,* 9, 1-13.

Coutinho, H. D. M., Costa, J. G. M., Siqueira, J. J. P., Lima, E. O. (2008). In vitro anti-staphylococcal activity of Hyptis martiusii Benth against methicillin-resistant Staphylococcus aureus-MRSA strains. *Rev. Bras. Farmacog.*, 18, 670-675.

Escobar, L. R. G. (2002). Eugenol: pharmacological and toxicological properties. Advantages and disadvantages of their use. *Rev. Cubana Estomatol.,* 39, 139 - 156.

Matias, E. F. F., Alves, E. F., Santos, B. S., Souza, C. E. S., Ferreira, J. V. A., Lavor, A. K. L. S., Figueredo, F. G., Lima, L. F., Santos, F. A. V., Peixoto, F. S. N., Colares, A. V., Boligon, A. A., Saraiva, R. A., Athayde, M. L., Rocha, J. B. T., Menezes, I. R. A., Coutinho, H. D. M., Costa, J. G. M. (2013). Biological Activities and Chemical Characterization of Cordia verbenacea DC.as Tool to Validate the Ethnobiological Usage. *ECAM,* 2013, 1-7.

Matias, E. F. F. (2010). Evaluation of the antibacterial activity and modulating antibiotic activity against aminoglycosides of poçar and non-polar extracts from Croton campestris A. (velame), Ocimum gratissimum L. (alfavaca) e Cordia verbanacea DC. (erva-baleeira). MSc. Thesis. Crato, Universidade Regional do Cariri, Brasil.

Matos, F. J. A. (1998). Live drugstore 3ª ed. Fortaleza, Ed. UFC, Brasil.

Murray, P. R., Rosenthal, K. S., Kobayashi, G. S. (2008). Microbiologia Médica. 4ª Ed. Guanabara Koogan, Rio de Janeiro, Brasil.

National Committee for Clinical Laboratory Standards – NCCLS. (2003). Methods for Dilution Antimicrobial Susceptibility Tests for Bacteria that grow aerobically, 6 ed. Wayne, NIH, USA.

Oliveira, F. Q., Gobira, B., Guimarâes, C., Batista, J., Barreto, M., Souza, M. (2007). Plants species indicated in odontology. *Rev. Bras. Farmacog.,* 17, 466-476.

Oliveira, R. A. G., Lima, E. O. L., Vieira, W. L., Freire, K. R. L., Trajano, V. N., Lima, I. O., Souza, E. L., Toledo, M. S., Silva, R. F. N. (2006). Study of the interference of essential oils on the activity of some antibiotic used clinically. *Rev. Bras. Farmacog.*, 16, 77-82.

Perreira, R. S., Sumita, T. C., Furlan, M. R., Jorge, A. O. C., Ueno, M. (2004). Antibacterial activity of essential oils on microorganisms isolated from urinary tract infection. Rev. *Saúde Publica*, 38, 326-328.

Salvat, A. A., Antonnacci, L., Fortunato, R. H., Suarez, E. Y. (2001). Screenng of some plants from northern Argentina for their antimicrobial activty. *Lett. Appl. Microbiol.*, 32, 293-297.

Sartoratto, A., Machado, A. L. M., Delarmelina, C., Filgueira, G. M., Duarte, M. C. T., Rehder, V. L. G. (2004). Composition and. antimicrobial activity of essential oils from aromatic plants used Brazil. *Braz. J. Microbiol.*, 35, 275-280.

Silva, J. G., Souza, I. A., Higino, J. S., Siqueira, J. J. P., Perreira, J. V., Pereira, M. S. V. (2007). Antimicrobial activity of the hydroalcoholic extract of *Anacardium occidentale* Linn. against multi-drug resistant strains of *Staphylococcus aureus*. *Rev. Bras. Farmacog.*, 17, 572-577.

Silva, M. G. F. (2011). In vitro Antioxidant and antimicrobial activities of essential oils and hydroalcohol extracts of manjerona (Origanum majorana L.) and manjericão (Ocimum basilicum L.). BSc. Thesis, Pato Branco, Universidade Tecnológica Federal do Paraná, Brasil.

Simon, G. (2004). Anti-staphylococcal plant natural products *Nat. Prod. Rep.*, 21, 263 - 277.

Sousa, V. C., Lorenzi, H. (2005). Systematic Botany: illutred guide to identification of of the angiospermae families of brazilian flora. 3ª. ed. Instituto plantarum, Nova Odessa, Brasil.

Vermelho, A. B., Bastos, M. C. F., Branquinha, M. H. S. (2007). General bacteriology. 1ª Ed. ed. Guanabara Koogan, Rio de Janeiro, Brasil.

Vieira, R. F., Grayer, R. J., Paton, A., Simon, J. E. (2001). Genetic diversity of Ocimum gratissimum L. based on volatile oil constituents, flavonoids and RAPD markers. *Biochem. Systemat. Ecol.*, 29, 287-304.

In: Essential Oils
Editor: Miranda Peters
ISBN: 978-1-63484-351-5

Chapter 5

CHEMICAL COMPOSITION AND ANTIMICROBIAL ACTIVITY OF *PLINIA JABOTICABA* (VELL.) KAUSEL ESSENTIAL OIL

Michele Debiasi Alberton[1,*], Gabriele Andressa Zatelli[1], Leonard de Vinci Kanda Kupa[1], Adrielli Tenfen[1], Diogo Alexandre Siebert[1], Juliana Bastos[2], Edesio Luis Simionatto[2] and Caio Maurício Mendes de Cordova[1]

[1]Department of Pharmaceutical Sciences, Universidade Regional de Blumenau, Blumenau, SC, Brazil
[2]Laboratory of Chromatography, Universidade Regional de Blumenau, Blumenau, SC, Brazil

ABSTRACT

The essential oil from the leaves of *Plinia jaboticaba* was analyzed by CG-MS. The sesquiterpenoids constituted the dominant fraction in this oil. The major compounds found in this oil were sesquiterpenes spathulenol (17.29%), γ-murolene (13.7%) and valencene (10.35%). The monoterpene fraction correspond to 9.28% of this essential oil, being 1,8-cineole (6.01%) the compound present in the higher quantity in this class. This oil was assayed to determine its antibacterial activity against cell-wall bacteria *Staphylococcus aureus*, *Escherichia coli* and *Pseudomonas aeruginosa*, and against no-cell-wall bacteria *Mycoplasma arginini*, *Mycoplasma mycoides* subsp. capri (MMC) and *Ureaplasma urealyticum*. The oil of *P. jaboticaba* showed antibacterial activity against all tested bacteria, which the most accentuated for no-cell-wall bacteria giving values of MIC = 750 μg/mL.

Keywords: antibacterial activity, chemical composition, essential oil, mollicutes, *Plinia jaboticaba*

[*] Corresponding author: michele@furb.br.

1. Introduction

Plinia is a genus with about 20 species ranging from Central America to southern Brazil [1]. *Plinia jaboticaba* (Vell.) Kausel (Myrtaceae), also recognized such *Myrciaria jaboticaba* is a cauliflorous tree native to Brazil, Argentina and Paraguay [2, 3]. Its fruit are used in folk medicine in the treatment of angina, gastroenteritis, erysipelas and asthma [4]. The bark is widely used by its astringency and also to threat diarrhea, as well as leaves are used to treat diarrhea [5-7]. Recently, some researchers have reported antioxidant activity of fruit extracts of *P. jaboticaba* [8, 9], and antimicrobial, antiulcer, antioxidant, anti-inflammatory and cytotoxic activity of extracts of species of this genus [10-17]. However, essential oil of leaves of *P. jaboticaba* has never been studied for any of the above mentioned purposes.

Essential oils are a complex mixture of compounds resulting from secondary metabolism of plants and present in many plants, including the *Plinia* genus [18]. Essential oils and the variety of chemical structures of their constituents are responsible for a wide range of biological activities, many of which are of increasing interest in the fields of human health. The main function of essential oils in plants, is a defense against environmental pathogens such fungus and bacteria [19]. The anti-infectious activity of essential oils is well known, and their importance in phytotherapy has been widely described. Their efficacy is due to a good antibacterial and antifungal activity for both human and plant pathogens; a wide range of activity; a lower insurgence of resistance, probably due to their mechanisms of action [20].

The relevant appearance of microbial resistance and the emergence of previously uncommon infections in humans have led the scientists into searching for new antimicrobial substances from various sources like natural products [21-23]. Because of the very few or almost absence reports on the composition of essential oil and biological activity of *P. jaboticaba* leaves, we decided to evaluate the chemical constituents and anti-bacterial activity of essential oil of *P. jaboticaba.*

2. Materials and Methods

2.1. Plant Material

Leaves of *Plinia jaboticaba* were collected in July 2011, in Blumenau (S 26° 53' 55; W 49° 04' 39), Santa Catarina State, Southern Brazil. Sample was identified by Prof. Dr. Marcos Eduardo Guerra Sobral, (Departamento de Ciências Naturais, Universidade Federal de São João Del-Rei/MG). Voucher specimen was deposited at Herbarium Roberto Miguel Klein (Blumenau/SC) at number 31234.

2.2. Extraction Procedure

Essential oil of leaves from *P. jaboticaba* was obtained by hydrodistillation for 4 h in a modified Clevenger-type apparatus. After extraction, the oils were dried with sodium sulfate and stored at a low temperature.

2.3. Analysis of Quantitative and Qualitative Composition of the Essential Oil

The analysis of components of essential oils from *P. jaboticaba* was performed by Gas Chromatography with Flame Ionization Detector GC-FID, Shimadzu 14b, column OV-5 (30 mx 0.25 mm id x 0.25 mm in film thickness). N_2 was used as carrier gas at a constant pressure of 80 kPa. The split ratio was 1/150 and injected with 0.2 μL of the pure oil. The detector temperature was 300 ° C and injector 250°C. The temperature of the column was 50°C (3 min) increase of 5°C/min to 270°C (8 min), resulting in a total time of analysis of 55 minutes. Gas Chromatography Coupled with Mass Spectrometry (GC-MS) was performed on GC VARIAN model CP 3800, with ionization by electron impact of 70 eV and source temperature of 170°C, in the same analytical conditions described for GC-FID. Individual components were identified by comparison of both their mass spectra and experimental retention indices (RI), which were calculated for all volatile constituents by using a homologous series (C8 to C19) recorded under the same operating conditions, and comparison with literature data [24]. The NIST 2008 (National Institute for Standards and Technology) was also used for comparison of mass spectra. The quantitative data were obtained electronic integration of the GC-FID peak areas.

2.4. Antibacterial Activity against Cell-Wall Bacteria

Microbiological Material

The bacterial strains were provided by the Laboratory of Clinical Microbiology of Universidade Regional de Blumenau (FURB) and acquired from The American Type Culture Collection (ATCC). Tests were carried out in triplicate with strains of *Staphylococcus aureus* (ATCC 25923), *Pseudomonas aeruginosa* (ATCC 27853), and *Escherichia coli* (ATCC 25922).

The identification of strains was confirmed by the use of biochemical profiles according to the recommendations of the Manual of Clinical Microbiology [25]. All organisms were maintained in brain-heart infusion (BHI) medium containing 30% (v/v) glycerol at 20°C. Before testing, the suspensions were transferred to trypticase soy agar supplemented with 5% sheep blood (Difco) and grown aerobically overnight at 35°C. The inocula were prepared by adjusting the turbidity of the suspension to match the 0.5 McFarland standard in saline solution (0.9%). Gentamycin was used as the reference antibiotic control.

Determination of Minimum Inhibitory Concentration (MIC)

The broth microdilution method was used to determine the MIC (minimum inhibitory concentration) of *P. jaboticaba* oil against the test organisms as recommended by the National Committee for Clinical Laboratory Standards [25]. This test was performed in sterile 96-microwell microplates. The oils were properly prepared and transferred to each microplate well in serial dilution of the original extract in DMSO (10%) from the initial concentration of 1000 μg/mL.

The inocula (10 μL) containing 5 x 10^5 CFU/mL of each microorganism were added to each well. A number of wells were reserved in each plate to test for sterility control (no inocula added), inocula viability (no extract added), positive control (gentamicin) and the DMSO 10% (negative control). Plates were aerobically incubated at 35°C. After incubation for 18-24 h, the plates were revealed with methanol solution of 2,3,5-triphenyl tetrazolium chloride (5 mg/mL) (10μL) and incubated at 35°C for 2 h. Bacterial growth in the wells was indicated by a red color, whereas clear wells indicated growth inhibition by the essential oil. MIC values were recorded as the lowest concentrations of essential oil showing clear wells. All the tests were performed in triplicate.

2.5. Antibacterial Activity against No-Cell-Wall Bacteria (Mollicutes)

Microbiological Material

The microorganisms used were reference strains of *Ureaplasma urealyticum* (ATCC 27618), *Mycoplasma arginini* (PG18, Institut Pasteur) and *Mycoplasma mycoides* subsp. capri (MMC) (PG3, NCTC 10137) were kept at -20ºC. *U. urealyticum* was inoculated into culture medium urea (U10), *M. arginini* in liquid arginine medium (MLA), and MMC in SP4 medium. All microorganisms were incubated at 36°C for 48 h under microaerophilic conditions.

Determination of Minimum Inhibitory Concentration (MIC)

The MIC was determined using the broth microdilution method [26]. The essential oil adjusted by at initial concentration of 2 mg/mL were diluted in U10 medium in the cases of *U. urealyticum*; MLA medium in the case of *M. arginini* and SP4 in the case of MMC [27].

First, the extracts and fractions were serially diluted by adding 100 μL of culture medium and added to 100 μL of the essential oil to obtain concentrations of 1000 to 7.8 μg/mL. To the essential oil were added either 100 μL of inoculum from *Mycoplasma spp.* or *U. urealyticum* culture in the log phase, containing 10^3 microorganisms/mL (determined by microtiter). The negative control was obtained from a serial dilution of the solvent (DMSO) and positive control was obtained from a serial dilution of the antibiotic (ciprofloxacin), without essential oil, and a serial dilution of the micro-organism culture without the addition of solvent or essential oil. Finally, to every cavity were added three drops of sterile petroleum jelly to insulate each cavity of the external environment and create a microaerophilic environment. The final concentration of DMSO in the cavities did not exceed 2%. The plates were incubated at 36°C for 48 h, the growth was observed from the color change of the culture medium, due to the presence of phenol red indicator. *Ureaplasma* species degrade urea to ammonia and alkalinize the pH, leading to a color change from yellow to red, therefore indicating the growth. *M. arginini* promote alkalinization of the medium in its growth by deamination of arginine in ammonia, and MMC use glucose as energy source, and growth is evidenced by the production of acids, turning indicator phenol red from red to yellow. The MIC was defined as the lowest extract or fraction concentration able to inhibit bacterial growth. All the tests were performed in triplicate.

3. Results and Discussion

3.1. Essential Oil Composition

After extraction, the yield (w/w) of the essential oil was calculated (0.13%). A total of 25 compounds were identified, accounting for 91.42% of the constituents by GC. Table 1 shows the qualitative and quantitative chemical composition of the sample studied. The sesquiterpenoids constituted the dominant fraction in this oil. No functionalized sesquiterpenes were the most frequently found (57.37%). However, the oxygenated sesquiterpene spathulenol was the compound present in the greatest concentration (17.29%). Other no-funcionalized sesquiterpenes like γ-murolene (13.7%), β-selinene (9.68%) and valencene (10.35%) were major compounds found in this oil. The monoterpene fraction correspond to 9.28% of this essential oil, being 1,8-cineole (6.01%) the compound present in the higher quantity in this class.

This is the first study of the chemical composition of essential oil of *P. jaboticaba* leaves. Previous analysis were performed only in the fruits of this species, and demonstrated the presence of 45 compounds, being 10 of these also found in this study: α-pinene, β-pinene, 1,8-cineol, α-copaene, β-elemene, caryophyllene-E, aromadrendene, α-humulene, spathulenol e selina-11-ene-4-α-ol [28].

Early studies performed with *Plinia* species show that the main fraction is composed by sesquiterpenes in several oils of this genus. Studies with other species known as *jaboticaba*, *P. trunciflora*, and *P. cauliflora* showed that spathulenol, the main compound found in the studied oil, was also found in significant amounts. It was found that several compounds of *P. jaboticaba* oil, as the γ-muurolene, α-pinene, α-terpineol, β-pinene, α-copaene, β-elemene, aromadendrene, allo-aromadendrene, α-humulene, β-selinene, γ-cadinene, 1,8-cineole, epi-α-cadinol, δ-elemene and cariophyllene (*E*) were commonly found in other species of *Plinia* genus. Moreover, some other compounds, as β-caryophyllene, bicyclogermacrene, caryophyllene oxide, globulol, epi-globulol, eudesmol, muurolol, α-bisabolol, α-bisabolol oxide B e α-bisabolol oxide A, that were also found in significant amounts in the oils of the mentioned species, were not found in the oil analysed in this study [3, 30, 31].

One can also notice that according to the studied oil, terpenes with caryophyllane and aromadendrane skeleton were specially characterized. These results are aligned with biosynthetic routes shown in previous studies on the chemical composition of essential oils of Myrtaceae, when the presence of caryophyllane, germacrane and aromadendrane compounds are a constant characteristic [31].

3.2. Antibacterial Activity

Table 2 shows the results of the antimicrobial assays with essential oil of *P. jaboticaba*. Aligianis et al. (2001) [32] proposed a classification for the antimicrobial activity of plant products, based on the MIC results as follows: strong inhibitors - MIC below 500 μg/mL; moderate inhibhitors - MIC between 600 and 1,500 μg/mL; weak inhibitors - MIC above 1,600 μg/mL. Thus, based on the MIC results obtained in the microplate assays, *P. jaboticaba* essential oil have interesting antibacterial potential, with MIC values below 1,600 μg/mL.

Table 1. Identified compounds in the essential oil of *Plinia jaboticaba* leaves

	Compound	Rt (min)[a]	RI (L)[b]	RI (E)[c]	%
1	α-pinene	7.233	932	929	1.66
2	β-pinene	8.866	979	970	0.88
3	1,8-cineole	11.300	1026	1026	6.01
4	α-terpineol	19.200	1186	1184	0.73
5	δ-elemene	26.500	1335	1333	1.88
6	α-copaene	28.316	1374	1371	3.28
7	β-elemene	29.050	1389	1387	1.08
8	Cariophyllene (E)	30.300	1417	1415	3.48
9	β-gurjunene	30.866	1431	1427	0.99
10	Aromadendrene	31.200	1439	1435	3.87
11	α-humulene	31.800	1452	1449	0.63
12	Aromadendrene (allo)	32.150	1458	1457	3.18
13	y-muurolene	33.133	1478	1479	13.78
14	β-selinene	33.366	1489	1484	9.68
15	Valencene	33.783	1496	1493	10.35
16	Guaiene-trans-β	34.083	1502	1500	0.56
17	γ-cadinene	34.483	1513	1510	0.94
18	Trans-calamenene	34.883	1522	1520	2.49
19	Germacrene (B)	36.283	1559	1554	1.18
20	Spathulenol	37.450	1577	1582	17.29
21	Viridiflorol	37.800	1592	1591	1.14
22	Himachalene oxide B	38.416	1615	1606	0.99
23	Cadinol (epi-α)	39.766	1638	1640	2.38
24	Selina-3,11, dien-6-α-ol	40.250	1642	1653	2.05
25	Selina-11-en-4-α-ol	40.500	1658	1659	0.92
Total identified					91.42
Total of monoterpenes					9.28
Hydrocarbon monoterpenes					2.54
Oxygenated monoterpenes					6.74
Total of sesquiterpenes					82.14
Hydrocarbon sesquiterpenes					57.37
Oxygenated sesquiterpenes					23.85

The identified constituents are listed in their order of elution from a nonpolar column OV-5.

[a] Rt (min) = Retention time (minutes).

[b] RI (L) = Retention index (literature) [24].

[c] RI (E) = Retention index (experimental).

The sample tested showed moderate activity against cell-wall bacteria *S. aureus*, *E. coli* and *P. aeruginosa*, with MIC values of 1000 μg/mL for these strains. This finding is interesting because these microorganisms are ethiologic agents commonly related to urinary tract infections, gastrointestinal disorders, skin infections and nosocomial infections. Other interesting aspect of this essential oil, is their activity against gram-positive and also against gram-negative bacteria. The great majority of plant extracts and essential oils are active against gram-positive microorganisms only. This phenomenon is attributed to significantly differences among gram-positive and negative cell-wall bacteria. The gram-negative bacteria presents a barrier to many substances, including antibiotics. This bacteria contains enzymes that able to break down foreign molecules, and efflux pumps that reduce the cellular levels of antibiotics [23].

Against no-cell-wall microorganisms *U. urealyticum* and *M. arginini*, the results also were classified such moderate. The *P. jaboticaba* essential oil showed MIC values = 750 µg/mL (Table 2). These results are important because Mycoplasma and Ureaplasma are obligate parasites of animals and humans that occasionally cause respiratory, urogenital and ocular disease. *U. urealyticum* is responsible for cases of non-gonococcal urethritis and bacterial vaginosis, and *M. arginini* is a species that infects sheep, being responsible for a severe respiratory disease in these animals and may cause an important zoonosis in humans [33]. MMC is a more ubiquitous pathogen in small ruminants causing mastitis, arthritis, keratitis, pneumonia and septicaemia and is also found as saprophyte in the ear canal [34]. Recently, the interest in plant extracts with anti-mycoplasma has increased, and methods have been developed to evaluate this activity [35].

Table 2. Antibacterial activity (MIC, µg/mL) of the essential oil of *P. jaboticaba* leaves

Bacterial Strains	EO[a]	Positive control[b]
Cell-wall bacteria		
Staphylococcus aureus	1000	5.0
Escherichia coli	1000	7.5
Pseudomonas aeruginosa	1000	1.0
No-cell-wall bacteria		
Ureaplasma urealyticum	750	1.5
Mycoplasma arginini	750	1.5
Mycoplasma mycoides subs. capri (MMC)	750	1.5

[a] EO = Essential oil.

[b] Gentamycin, for cell-wall bacteria; ciprofloxacin, for no-cell-wall bacteria.

Furthermore, the results obtained in these experiments, suggests that *P. jaboticaba* essential oil exerts their effects by one mechanism not related to the cell wall disruption. This statement is possible because the *P. jaboticaba* essential oil maintained the same pattern of antibacterial action observed against cell-wall bacteria (*S. aureus*, *E. coli* and *P. aeruginosa*) and no-cell-wall bacteria such Mycoplasmas (*U. urealyticum* and *M. arginini*). Mycoplasmas are the smallest no-cell-wall free-living prokaryotes [36]. However, these microorganisms have plasma membranes rich in cholesterol. According to Trombetta et al. (2005) [36], the mechanism of action of terpenes is associated to a damage of the plasma membrane stability. An important characteristic of essential oils and their components, like terpenes, is their hydrophobicity, which enables them to partition in the lipids of the bacterial cell membrane and mitochondria, disturbing the structures and rendering them more permeable [38].

Studies with liposome model systems confirmed that cyclic terpene hydrocarbons, the major compounds found in this oil, accumulate in the membrane, which causes a loss of membrane integrity and dissipation of the proton motive force [39]. Considering the large number of different groups of chemical compounds present in essential oil, it is most likely that their antibacterial activity is not attributable to one specific mechanism but that there are several targets in the cell. Leakage of ions and other cell contents can then occur. Although a certain amount of leakage from bacterial cells may be tolerated without loss of viability, extensive loss of cell contents or the exit of critical molecules and ions will lead to death. Components of essential oil also appear to act on cell proteins embedded in the cytoplasmic membrane, like enzymes such as ATPases [38]. Previous studies have shown *in vitro* activities of essential oil of Myrtaceae family against mycoplasmas [40].

CONCLUSION

At the end of this study we may conclude that the tested essential oil has the composition in accordance to the biosynthetic pathways of *Plinia* genus. It is also promising alternative, mostly when it comes to the treatment of infections caused by bacteria of mollicutes class. More detailed studies using isolated compounds are necessary to evaluate if there is synergism between constituents of this oil, or only one compound is the responsible for the observed antibacterial activity.

ACKNOWLEDGMENTS

The authors thank to CNPq, FAPESC and FURB for financial support.

REFERENCES

[1] Landrum, L. R., Kawasaki, M. L. The genera of Myrtaceae in Brazil - an illustrated synoptic treatment and identification keys. *Brittonia,* 1997, 49, 508-536.

[2] Magalhães, M. M., Barros, R. S., Finger, F. L. Changes in structural carbohydrates in developing fruit of *Myrciaria jaboticaba. Scientia Horticulturae,* 1996, 66, 17-22.

[3] Duarte, A. R., Santos, S. C., Seraphin, J. C., Ferri, P. H. Environmental influence on phenols and essential oils of *Myrciaria cauliflora* leaves. *Journal of the Brazilian Chemical Society,* 2010, 21, 1672-1680.

[4] Azeredo, E. H. Comparison and color efficiency in a model trap with ethanolic extract and fins to catching insect associated with pests in *Myrciaria jabuticaba* (Berg) (Myrtaceae). *Journal of Rural University,* 2006, 26, 54-67.

[5] Cruz, A. V. M., Kaplan, M. A. C. Medicinal use of species from Myrtaceae and Melastomataceae families in Brazil. *Forest and environment,* 2004, 11, 47-52.

[6] Lorenzi, H. *Árvores Brasileiras:* Manual for Identification and Cultivation of Arboreal native plants in Brazil. New Odessa: Plantarum Institute, 2000.

[7] Albuquerque, U. P., Medeiros, P. M., Almeida, A. L. S., Monteiro, J. M., Lins Neto, E. M. F., Melo, J. G., Santos, J. P. Medicinal plants of the caatinga (semi-arid) vegetation of NE Brazil: a quantitative approach. *Journal of Ethnopharmacology*, 2007, 114, 325-354.

[8] Vieites, R. L., Daiuto, E. R., Moraes, M. R., Neves, L. C., Carvalho, L. R. Physical-chemical, biochemical and functional characterization of jabuticaba stored at different temperatures. *Brazilian Journal of Fruticulture*, 2011, 33, 362-375.

[9] Leite, A. V., Malta, L. G., Riccio, M. F., Eberlin, M. N., Pastore, G. M., Maróstica Júnior, M. R. Antioxidant potential of rat plasma by administration of freeze-dried *jaboticaba* peel (*Myrciaria jaboticaba* Vell Berg). *Journal of Agricultural and Food Chemistry,* 2011, 59, 2277-2283.

[10] Ishikawa, T., Kato, E. T. M., Yoshida, M., Kaneko, T. M. Morphoanatomic aspects and phytochemical screening of *Plinia edulis* (Vell.) Sobral (Myrtaceae). *Revista Brasileira de Ciências Farmacêuticas,* 2008, 44, 515-520.

[11] Ishikawa, T., Donatini, R. S., Diaz, I. E. C., Yoshida, M., Bacchi, E. M., Kato, ETM. Evaluation of gastroprotective activity of *Plinia edulis* (Vell.) Sobral (Myrtaceae) leaves in rats. *Journal of Ethnopharmacolo-gy,* 2008, 118, 527-529.

[12] Einbond, L. S., Reynertson, K. A., Luo, X., Basile, M. J., Kennelly, E. J. Anthocyanin antioxidants from edible fruits. *Food Chemistry,* 2004, 84, 23-28.

[13] Reynertson, K. A., Yang, H., Jiang, B. B., Basile, M. J., Kennelly, E. J. Quantitative analysis of antiradical phenolic constituents from fourteen edible Myrtaceae fruits. *Food Chemistry,* 2008, 109, 883-890.

[14] Rufino, M. S. M., Alves, R. E., Brito, E. S., Pérez-Jiménez, J., Saura-Calixto, F., Mancini-Filho, J. Bioactive compounds and antioxidant capacities of 18 non-traditional tropical fruits from Brazil. *Food Chemistry,* 2010, 121, 996-1002.

[15] Macedo-Costa, M. R., Diniz, D. N., Carvalho, M. S. V. P., Pereira, J. V., Higino, J. S. Eficácia do extrato de *Myrciaria cauliflora* (Mart.) O. Berg. (jabuticabeira) sobre bactérias orais. *Brazilian Journal of Pharmacogno-sy,* 2009, 19, 565-571.

[16] Reynertson, K. A., Basile, M. J., Kennelly, E. J. Antioxidant potential of seven myrtaceous fruits. *Ethnobotany Research and Applications.,* 2005, 5, 25-36.

[17] Mohanty, S., Cock, I. Evaluation of the antibacterial activity and toxicity of *Myrciaria cauliflora* methanolic leaf and fruit extracts. *The Internet Journal of Microbiology.,* 2009, 7, n 2.

[18] Delamare, A. P. L., Moschen-Pistorello, O. I. T., Artico, L., Atti-Serafini, L., Echeverrigaray, S. Antibacterial activity of the essential oils of *Salvia officinalis* L. and *Salvia triloba* L. cultivated in South Brazil. *Food Chemistry.,* 2007, 100, 603-608.

[19] Oliveira, R. A. G., Lima, E. O., Viera, W. L., Freire, K. R. L., Trajano, V. N., Lima, I. O., Souza, E. L., Toledo, M. S., Silva-Filho, R. N. Interference of essential oils on the activity of some antibiotics with clinical uses. *Brazilian Journal of Pharmacognosy,* 2006, 16, 77-82.

[20] Furneri, P. M., Mondello, L., Mandalari, G., Paolino, D., Dugo, P., Garozzo, A., Bisignano, G. *In vitro* antimycoplasmal activity of *Citrus bergamia* essential oil and its major components. *European Journal of Medicinal Chemistry,* 2012, 52, 66-69.

[21] Zampini, I. C., Vattuone, M. A., Isla, M. I. Antibacterial activity of *Zuccagnia punctata* Cav. Ethanolic extracts. *Journal of Ethnopharmacology,* 2005, 102, 450-456.

[22] Pasqua, R., Feo, V., Villani, F., Mauriello, G. *In vitro* antimicrobial activity of essential oils from Mediterrenean Apiaceae, Verbanaceae and Limeaceae against foodborns pathogens and spoilage bacteria. *Annals of Microbiology,* 2005, 55, 132-143.

[23] Magina, M. D. A., Dalmarco, E. M., Wisniewski Junior, A., Simionatto, E. L., Dalmarco, J. B., Pizzolatti, M. G., Brighente, I. M. C. Chemical composition and antibacterial activity of essential oils of Eugenia species. *Journal of Natural Medicines,* 2009, 63, 345-350.

[24] Adams, R. P. *Identification of essential oil compounds by gas chromatography/mass spectroscopy*. Illinois: Allured publishing Co. Carol Stream, 2007.

[25] CLSI. 2012. *Performance Standards for Antimicrobial Susceptibility Testing: Twenty-Second Informational Supplement*. Wayne, PA, Clinical and Laboratory Standards Institute, (M100-S22/v. 32, n. 3).

[26] Bebear, C., Robertson, J. *Determination of the minimal inhibitory concentration*. In: Tully J. G., Razin S. (ed). Molecular and Diagnostic Procedures in Mycoplasmology. vol. II. San Diego: Academic Press, 1996, 189-197.

[27] Velleca, W. M., Bird, B. R., Forrester, F. T. *Complete media for growth of mycoplasmas and ureaplasmas from the urogenital tract: arginine broth medium for growth of M. hominis*. In: Velleca W. M., Bird B. R., Forrester F. T. Laboratory diagnosis of Mycoplasma infections: course 8226-C. Atlanta: U.S. Department of Health, Educationand Welfare, C.D.C., 1979.

[28] Plagemann, I., Krings, U., Berger, R. G., Marostica Junior, M. R. Volatile constituents of Jabuticaba (*Myrciaria cauliflora* (Vell) O. Berg). *Journal of Essencial Oil Research,* 2012, 24, 45-51.

[29] Apel, M. A., Sobral, M., Zuanazzi, J. A., Henriques, A. T. Essential oil composition of four *Plinia* species (Myrtaceae). *Flavour and Fragrance Journal,* 2006, 21, 565-567.

[30] Vila, R., Santana, A. I., Pérez-Rosés, V. A., Castelli, S. M., Zacchino, S., Gupta, M. P., Cañigueral, S. Composition and biological activity of the essential oil from leaves of *Plinia cerrocampanensis*, a new source of α-bisabolol. *Bioresource Technology,* 2010, 101, 2510-2514.

[31] Fortes, G. C., Naves, S. S., Godoi, F. F. F., Duarte, A. R., Ferri, P. H., Santos, S. C. Assessment of maturity index in jabuticaba fruit by the evaluation of phenolic compounds, essencial oil componentes, sugar content and total acidity. *American Journal of Food Technology,* 2011, 6, 974-984.

[32] Aligiannis, N., Kalpotzakis, E., Mitaku, S., Chinou, I. B. Composition and antimicrobial activity of the essential oil of two *Origanum* species. *Journal of Agricultural and Food Chemistry,* 2001, 40, 4168-41.

[33] Cordova, S. M., Benfatti, C. S., Magina, M. D. A., Guedes, A., Cordova, C. M. M. Evaluation of antimicrobial activity of extracts from brazilian native flora against *Mycoplasma arginini, M. hominis* and *Ureaplasma urealyticum. Brazilian Journal of Clinical Analysis,* 2010, 42, 241-244.

[34] Thiaucourt, F., Manso-Silvan, L., Salah, W., Barbe, V., Vacherie, B., Jacob, D., Breton, M., Dupuy, V., Lomenech, A. M., Blanchard, A., Sirand-Pugnet, P. *Mycoplasma mycoides*, from "*mycoides Small Colony*" to "*capri*": a microevolutionary perspective. *BMC Genomics,* 2011, 16, 112-114.

[35] Muraina, I. A., Picard, J., Eloff, J. N. Development of a reproducible method to determine minimum inhibitory concentration (MIC) of plant extract against a slow-growing mycoplasmas organism. *Phytomedicine,* 2009, 16, 262-264.

[36] Razin, S., Yogev, D., Naot, Y. Molecular biology and pathogenicity of mycoplasmas. *Microbiology and Molecular Biology Reviews,* 1998, 62, 1094-1156.

[37] Trombetta, D., Castelli, F., Sarpietro, M. G., Venuti, V., Cristani, M., Daniele, C., Saija, A., Mazzanti, G., Bisignano, G. Mechanism of antibacterial action of monoterpenes. *Antimicrobial Agents and Chemotherapy,* 2005, 49, 2474-2478.

[38] Burt, S. Essential oils: their antibacterial properties and potential applications in foods - a review. *International Journal of Food Microbiology,* 2004, 94, 223-253.

[39] Sikkema, J., De Bont, J. A. M., Poolman, B. Mechanisms of membrane toxicity of hydrocarbons, *Microbiology Review.,* 1995, 59, 201-222.

[40] Furneri, P. M., Paolino, D., Saija, A., Marino, A., Bisignano, G. *In vitro* antimycoplasmal activity of *Melaleuca alternifolia* essential oil. *Journal of Antimicrobial Chemotherapy,* 2006, 58, 706-707.

In: Essential Oils
Editor: Miranda Peters

ISBN: 978-1-63484-351-5

Chapter 6

ESSENTIAL OILS AS POTENTIAL ANTIOXIDANT AGENTS IN THE TREATMENT OF OXIDATIVE STRESS-RELATED NEURODEGENERATIVE DISEASES: RESULTS ON *IN VIVO* AND *IN VITRO* MODELS

Carlos Fernández-Moriano*, Elena González-Burgos*
and M. Pilar. Gómez-Serranillos
Department of Pharmacology, Faculty of Pharmacy,
University Complutense of Madrid, Spain

Oxidative stress is defined as an imbalance between pro-oxidant and antioxidant species within the intracellular environment, where an overproduction of reactive oxygen species (ROS) may cause oxidative damage to cell structures, such as DNA, functional proteins or cell membrane lipids; the perpetuation of that situation finally leads to cell death. Extensive evidence demonstrate the key role of oxidative stress in the etiopathogenesis of neurodegenerative disorders, including Alzheimer´s and Parkinson´s diseases among others, which are characterized by the loss and dysfunction of nervous cells.

The use of natural exogenous antioxidants capable to counteract the oxidative damage via diverse mechanisms, such as scavenging of reactive species, potentiation of endogenous enzymatic defense or amelioration of mitochondrial dysfunction, attracts increasing research and emerges as a promising therapeutic tool. In this context, several natural products are currently being investigated as potential neuroprotector agents.

Essential oils are concentrated liquids with pleasant odors and composed of complex mixtures of volatile compounds that can be extracted from several plant organs by distillation processes. They are a good source of several bioactive compounds with pharmacological interest, and have been largely reported for antimicrobial, anti-inflammatory, anticancer and antioxidative properties. Some essential oils have been used in clinical medicine with different purposes, varying from hypnotic induction in sleep disturbances to the treatment of respiratory ailments.

* These authors contributed equally to this work.

The current chapter offers a general overview of the implications of oxidative stress in the neurodegeneration of Alzheimer´s and Parkinson´s diseases, with special focus on the antioxidant and neuroprotective properties demonstrated for essential oils. Results obtained in several neurodegeneration models are collected in this review and their therapeutic potential discussed.

1. Oxidative Stress

Oxidative stress is a redox system imbalance that occurs when the production of reactive oxygen species (ROS) overpass the capacity of the enzymatic and non-enzymatic antioxidant defense system to neutralize and eliminate them. Under oxidative stress conditions, the excess of ROS induce alteration of macromolecules (lipids, proteins, DNA and RNA), leading to an oxidative cellular damage and even to cellular death through necrosis or apoptosis pathways (Figure 1) [1, 2].

Increased ROS can attack any of the carbon-carbon double bonds of poly-unsaturated fatty acids (PUFAs) at the cellular membrane components through self-propagation chain-reaction mechanisms, resulting in lipid peroxidation and in a modification of the lipid bilayer structure and properties (i.e., fluidity, thickness and permeability). During the oxidation of lipids, it is produced a wide variety of end-products including 4-hydroxinonenal (4-HNE) and malonyldialdehyde (MDA) which are the most physiologically active lipid peroxides; 4-hydroxinonenal (4-HNE) is the most toxic product of lipid peroxidation and malonyldialdehyde is considered the major mutagenic and carcinogenic product [3-5].

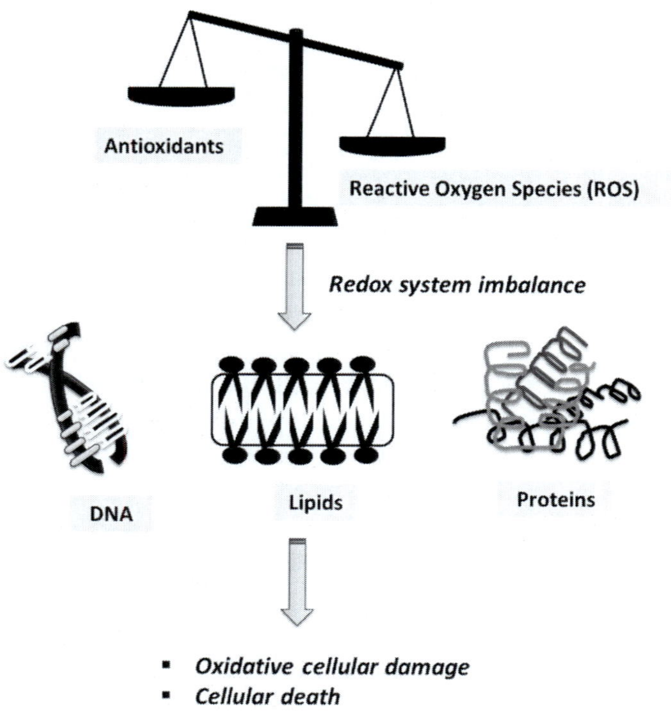

Figure 1. Scheme summarizing the cellular effects of oxidative stress.

Moreover, proteins can be directly modified by the oxidative action of ROS; these reactive molecules can oxidize protein backbones and amino acid residues, specially cysteine and methionine, resulting in protein-protein cross-linkages and protein fragmentation. In addition to ROS action, proteins can be indirectly modified by reaction with reactive sugars via glycation or glycoxidation reactions and by reaction with secondary by-products of lipid peroxidation such as the afore-mentioned 4-HNE and MDA. These oxidative modifications in proteins alter both structure and function (i.e., solubility, conformation, susceptibility to proteolysis, and enzyme activities). For example, the oxidation of prolyl residues leads to the formation of the 2-pyrrolidone and the oxidation of lysyl residues gives to α-aminoadipic semialdehyde products. As another example, the superoxide anion radical oxidize the [4Fe-4S] cluster of the mitochondrial aconitase enzyme, causing an inactivation of the Krebs cycle step mediated by this enzyme [6-8].

Furthermore, ROS can oxidize and cause damage to the RNA and DNA molecules. For the DNA, the most frequent lesions are site of base loss (apurinic/apyrimidinic (AP) abasic sites), base modifications (oxidized purines and pyrimidines among which highlight 8-Oxo-2'-deoxyguanosine (8-oxo-dG) and 2,6- diamino-4-hydroxy-5-formamidopyrimidine) and, single strand breaks (SSBs) and double strand breaks (DSB) which are related to single genes mutations and chromosome aberrations [9, 10]. On the other hand, RNA has some characteristics that make it more vulnerable to the harmful effects of ROS: the ribonucleic acid is mainly single-stranded which leaves it more easily accessible to ROS, it shows a subcellular distribution and it has not repair mechanisms for oxidative damage. The most common oxidized base in RNA is 8-hydroxyguanosine (8-OHG). Oxidative damage to RNA modifies its structure and its function as evidenced in disturbance of the translational process and impairment of protein synthesis [11, 12].

Reactive Oxygen Species (ROS)

Reactive oxygen species (ROS) are chemically reactive molecules derived from molecular oxygen which comprise both free radicals and non-free radicals. Free radicals, like superoxide anion, hydroxyl radical (•OH), peroxyl radicals and alkoxy radicals, among others, are chemically characterized by the presence of unpaired electrons in their outer molecular orbitals. On the other hand, non-free radicals, like singlet oxygen, hydrogen peroxide, peroxynitrite, hypochlorous acid and hypobromous acid, among others, do not contain any unpaired electron but are reactive species that can be converted into free radicals [13, 14].

ROS can be formed by several different mechanisms: (1) exposure to ionizing radiation (UV light and X-rays and gamma rays), (2) intracellular organelle sources: mitochondria (the major organelle responsible for the production of ROS), endoplasmic reticulum and peroxisomes, (3) pesticides, air and water pollution, food, smoking, etc., (4) inflammatory cells such as macrophages, eosinophils and neutrophils, (5) during metal catalyzed oxidation reaction such as Fenton and Haber-Weiss reactions and (6) enzymatic ROS production reactions: NADPH oxidase, lipoxygenases, xanthine oxidase, cyclo-oxygenases, nitric oxide synthase and cytochrome P450, among other examples [15, 16].

In physiological conditions, ROS have beneficial effects. These reactive species kill different pathogens in humans by causing oxidative damage or by stimulating pathogen elimination. Moreover, they are necessary for the biosynthesis of the thyroid hormone and they also regulate protein phosphorylation, ion channels (i.e., calcium and potassium channels), and transcription factors (i.e., IL-1β, IL-10, IL-18 and TNF-α). Furthermore, at cardiovascular level, ROS regulate vascular homeostasis and they also act as second messengers modulating mitogen-activated protein kinases, protein tyrosine kinases and protein tyrosine phosphatases [17-19].

However, when produced in excess, ROS may cause a redox imbalance (oxidative stress) and damage to cellular biomolecules leading to structural and functional alterations and even to cell death. Excessive ROS generation has been involved in the pathogenesis of several acute and chronic disease processes including neurodegenerative disorders such as Parkinson´s and Alzheimer´s diseases (in which we will focus on this chapter), cancer and atherosclerosis, among others [20-22].

Antioxidants

Antioxidants are compounds that protect cells from ROS-induced damage by scavenging them, by binding metal ions or/and by up-regulating endogenous antioxidant enzymes. There are endogenous antioxidant compounds (enzymatic and non- enzymatic) and exogenous antioxidant compounds with natural, semisynthetic or synthetic origin [23, 24].

Among enzymatic and endogenous antioxidant compounds, one may highlight the enzymes catalase, superoxide dismutase, glutathione peroxidase and glutathione reductase. Catalase (CAT) is an antioxidant enzyme located mainly in peroxisomes which efficiently converts hydrogen peroxide into water and molecular oxygen. Superoxide dismutase (SOD) catalyzes the transformation of superoxide anion into hydrogen peroxide and molecular oxygen, and three isoforms of SOD enzyme have been described: CuZnSOD (SOD1) located in cytosol, MnSOD (SOD2) located in mitochondria and EC-SOD (SOD3) which is extracellular. Glutathione peroxidase (GPx) enzymes (selenium-dependent and selenium independent) catalyze the reduction of hydroperoxides using glutathione (GSH). Glutathione reductase (GR), located in cytoplasm and mitochondria, converts oxidized glutathione (GSSG) into two molecules of reduced glutathione using NADPH as cofactor [25, 26].

Among non-enzymatic endogenous antioxidants, glutathione deserves to be highlighted. This antioxidant is composed of three different amino acids: L-cysteine, L-glutamic acid, and glycine. There are two forms for glutathione: oxidized glutathione (GSSG) and reduced glutathione (GSH). The reduced glutathione is the active form and it exerts its action through the sulfhydryl group (SH) of cysteine which serves as a proton donor. This antioxidant has the ability to scavenge ROS and RNS and, moreover, it acts as cofactor of different enzymes including glutathione peroxidase, as previously commented [27, 28].

Under oxidative stress situation, endogenous antioxidant compounds are insufficient to defend against excessive ROS production, being necessary to supplement with exogenous antioxidants found mainly in plants and different food (fruits, vegetables…). In this chapter, we will focus in the role of essential oils as antioxidants [29, 30].

2. Essential Oils: An Overview

Essential oils are aromatic and volatile liquids containing mixtures of various classes of organic compounds (mostly terpenes, terpenoids and phenols, and in a lesser extent alcohols, aldehydes, ketones, esters, lactones and coumarins) (Figure 2). Essential oils are responsible for the sensorial appealing of odour and flavour of plants and, for these organoleptic properties they are employed to make perfumes and flavorings. Moreover, plant essential oils possess different biological activities including antimicrobial, antioxidant, anti-inflammatory and anti-cancer, among others, which will be discussed in the final part of this section [31, 32].

Essential oils are synthetized and storage in secretory glandules of different plant organs including the flowers [like that of jasmine (*Jasminum officinale*), lavender (*Lavandula angustifolia*) and rosemary (*Rosmarinus officinalis*)], in the fruits [such as those of the orange (*Citrus sinensis*), lemon (*Citrus lemon*) and tangerine (*Citrus tangerine*)], in the leaves [like in peppermint (*Mentha piperita*), eucalyptus (*Eucalyptus citriodora*) and cypress (*Cupressus sempervirens*)], in the wood [as in the case of the sandal (*Santalum album*) and cedar (*Juniperus virginiana*)], in the seeds [cardamom (*Elettaria cardamomum*) and coriander (*Coriandrum sativum*)], in the bark [such as that of cinnamon (*Cinnamomum zeylanicum*)] and in the rhizomes and roots [i.e., the essential oil of Canada wild ginger (*Asarum canadense*)] [33-36].

Nowadays, essential oils from about 400 species belonging to 67 plant families have been reported to be commercially available, among which we could highlight those from Lauraceae, Myrtaceae, Umbelliferae, Labiatae and Compositae families [37].

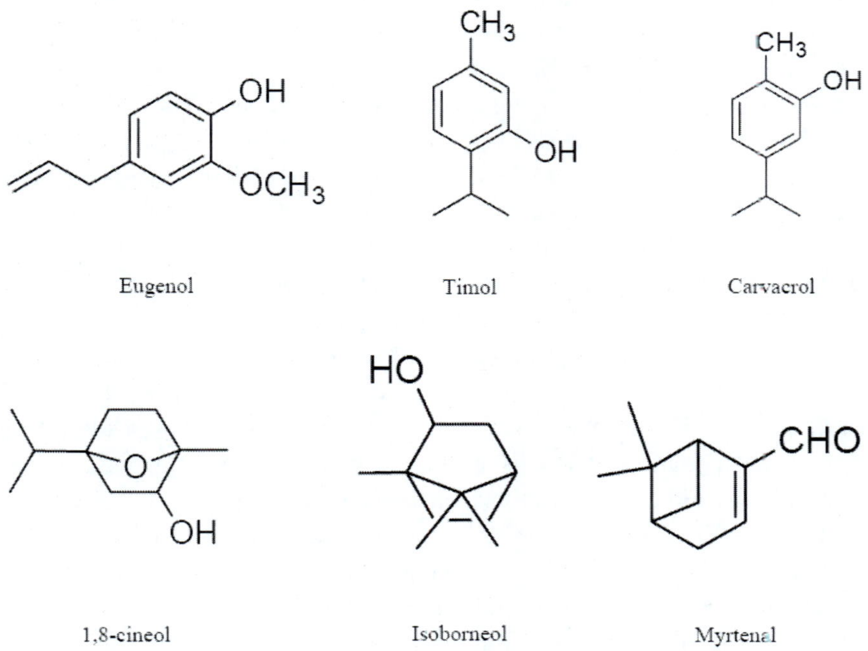

Figure 2. Examples of chemical structures of compounds containing in essential oils.

The amount and the chemical composition of the essential oils are influenced by many different factors including soil and weather conditions, harvest time, agronomic management, storage conditions and genetic make up of the plant, among others. As example, Verma et al. (2011) demonstrated that the essential oils obtained from *Thymus serpyllum* plants harvested in the summer give better essential oil yield than *Thymus serpyllum* plants harvested in other seasons [38]. As another example, the essential oils from the leaves and flowers of *Callistemon citrinus* were different in terpenoic compounds quantity and types; the essential oil obtained from flowers was rich in 1,8-cineole and α-pinene and the essential oil from leaves had high content of α-pinene, limonene and α-terpineol [39]. Moreover, the yield of the oil and the concentration of the bioactive compounds β-caryophyllene, limonene and terpinen-4-ol obtained from *Myrocarpus frondosus* essential oil were found to be directly proportional to temperature and radiation [40]. Furthermore, Choi and Sawamura revealed that both time and temperature of storage cause compositional changes in the essential oil of *Citrus tamurana*; the higher the time and temperature during storage, the higher the content in monoterpene alcohols and ketones, in the monoterpene hydrocarbon p-cymene and in the sesquiterpene hydrocarbons (-)-cedrene, γ-elemene, and α-humullene, and the lower the content in the monoterpene hydrocarbons myrcene, gamma-terpinene, and terpinolene and in the sesquiterpene hydrocarbon trans-beta-farnesene, among other examples [41].

Essential oils can be extracted by classical methods (i.e., hydrodistillation, water steam entrainment, organic solvent extraction and cold pressing) and by innovative techniques (i.e., supercritical fluid extraction, subcritical extraction liquids, extraction with subcritical CO_2, ultrasound assisted extraction and microwave assisted extraction) [42].

Traditionally, essential oils have been widely employed for its antiseptic properties (effective antibacterial, antivirus and antifungical action) and nowadays, these antimicrobial activities are used for food and pharmaceutical applications [43]. For example, the essential oils from tea tree, white thyme and oregano have demonstrated to be effective antibacterial products against *Salmonella typhimurium*, *Salmonella enteritidis*, *Escherichia coli*, *Staphylococcus aureus* and *Enterococcus faecalis* in animal feed, food, and sanitizers [44]. Moreover, oregano essential oil and its major component carvacrol are active against against the nonenveloped murine norovirus (MNV) acting on its viral capsid and RNA [45]. Furthermore, the essential oils obtained from coriander and thyme possess potential antifungical and antibacterial properties for gynecological treatment of those infections caused by *Escherichia coli*, *Staphylococcus aureus* and *Candida albicans* [46].

Essential oils have also demonstrated anticarcinogenic activity. The essential oil obtained from the seeds of *Descurainia Sophia* has resulted to possess cytotoxic activity on the cancerous cell lines MCF-7 (human breast adenocarcinoma cell line) and HeLa (human cervical cancer cells) [47]. Moreover, carvacrol, a major compound presented in the essential oil of thyme, rosemary and oregano, is an effective anticancer compound against colon (HCT116 and LoVo cell lines) and liver cancer (human hepatoma HepG2 cells) via inducing apoptosis pathways [48, 49]. Furthermore, Russo et al. (2015) evaluated and compared the anticancer properties of *Salvia verbenaca* essential oils of plants cultivated and growing in natural sites; results demonstrated that both essential oil types inhibited the growth of the human melanoma cancer cell line M14 through apoptotic mechanisms and this anticancer properties were higher for the essential oil obtained from cultivated plants (which contains hexahydrofarnesyl acetone as major compound) than for the essential oil from wild plants (which contains hexadecanoic acid as major compound) [50]. As a final example, the essential oil of *Artemisia vulgaris* and the essential oil of *Cryptomeria japonica* possess anticancer properties against HL-60 leukemic cell line and human oral epidermoid carcinoma cells, respectively, by inducing mitochondria-dependent apoptosis [51, 52].

Essential oils have been also investigated for its potential benefits as hypoglycemic agents. The essential oil from the leaf of *Hoslundia opposite* reduced fasting blood glucose (FBG) levels and glucose concentrations in rats as well as it reversed the cytoarchitectural distortion of pancreatic β-cells of diabetic rats [53]. Moreover, the essential oil obtained from the leaves of *Pinus koraiensis* has shown to possess hypoglycemic potential in streptozotocin-treated mice and HIT-T15 pancreatic β cells and this property is attributed to its capacity to inhibite reactive oxygen species (ROS), endothelial NO synthase (eNOS) and vascular endothelial growth factor (VEGF) [54]. Furthermore, Lee et al. (2013) demonstrated the hypoglycemic and pancreatic-protective effect of leaf essential oil from indigenous cinnamon (*Cinnamomum osmophloeum* Kanehira) in diabetic rats induced with streptozotocin via anti-inflammatory and anti-oxidative stress pathways [55].

Essential oils are also effective antioxidants. As examples, the essential oil obtained from the rhizomes of *Zingiber officinale* has scavenging properties as evidenced by 1,1-diphenyl-2-picryl-hydrazyl (DPPH), 2,2-azinobis(3-ethylbenzothiazoline-6-sulfonate) (ABTS) and hydroxyl radical scavenging tests, and it also increases the activity of superoxide dismutase (SOD), catalase (CAT) and glutathione peroxidase (GPx) enzymes [56]. In another study, the essential oil obtained from the leaves of *Psidium cattleianum* demonstrated to have antioxidant potential by scavenging free radicals (DPPH, ABTS and ferric ion reducing antioxidant power (FRAP)) and by inhibiting lipid peroxidation and increasing catalase enzyme activity [57].

In the present chapter, we will focus on the role of the essential oils and its main components as antioxidants and neuroprotectors in the prevention and treatment of Alzheimer´s disease and Parkinson´s disease.

3. Essential Oils as Neuroprotective Agents in Alzheimer´s Disease Models

Alzheimer's disease (AD) is the most prevalent age-related neurodegenerative disorder with characteristic clinical and pathological features associated to loss of neurons in certain brain areas, including cholinergic and glutamatergic systems; this normally provokes impairment of memory, cognition dysfunction, pyschobehavioural disturbances and deficits in activities of daily living, and eventually leads to death. In 2010, approximately 35 million people worldwide were suffering from AD, and this number is believed to increase to 66 million by 2030. To date, limited therapeutic drugs are available, and they mostly target the associated neuronal systems, namely acetylcholinesterase inhibitors and N-methyl-D-aspartate (NMDA) glutamate channel blockers [58, 59].

Concerning the etiopathology of the disease, oxidative stress plays a central role in the initiation and progression of AD; the brain is particularly vulnerable to oxidative damage because of its elevated oxygen utilization rate, the high content of polyunsaturated lipids (which are susceptible to lipid peroxidation), the accumulation of transition metals (which are capable of catalyzing the formation of ROS) and the relative paucity of cellular antioxidants.

Intracellular and extracellular amyloid beta-protein (Aβ) accumulation and deposition are major features of AD and are also closely related to the excessive oxidative stress in this disease; this neurotoxic protein inhibits the electron transport chain in mitochondria, decreases the respiratory rate, induces the release of ROS and may also cause neurotoxicity through the direct production of ROS via its interaction with transition metals and lipid membranes [60, 61].

As previously exposed, to combat the cytotoxicity of ROS, cells are endowed with a variety of antioxidant defense mechanisms, including protective enzymes as well as free-radical scavengers. However, under the redox imbalance present in Alzheimer´s disease and due to the lack of effective treatment, there is a need for the search of novel antioxidant compounds with neuroprotective potential that are able to counteract this situation. Natural antioxidants derived from a multitude of plants have demonstrated to display promising cytoprotective properties and constitute an interesting group of compounds that can be used as drug candidates [62].

Several plant essential oils containing bioactive components are of interest as plausible therapeutic options for neuroprotection in Alzheimer´s disease, mainly via antioxidative mechanisms. Due to the fact that the cholinergic system is probably the most important one for understanding AD pathology and also the main target for the currently available drugs, several natural products have been investigated regarding their capacity to inhibit the enzyme acetylcholinesterase; herein, it is therefore mandatory to highlight the essential oils that have displayed this activity in several models. Thus, throughout this part of the chapter, we proceed to report all the available information on the study of essential oils as potential protective agents in both *in vitro* and *in vivo* models of Alzheimer´s disease, with special focus on the most recent discoveries (Table 1).

Aromatherapy experientially classifies the effect of the scent through the essential oil extracted from the plant and is traditionally used in many fields. It has been suggested that aromatherapy may bring about some feeling of relief and the ability to act on outside influences in senile dementia. There are only few reports of the effects of aromatherapy on cognitive functional disorder, which is the central feature of senile dementia and AD. The available scientific data indicate that the olfactory tract and the generation of nerve impulses into olfactory bulb are closely related, structurally and functionally, to the hippocampus. As hippocampus is involved in learning and memory functions, olfaction-related memory can directly participate in distinguishing, coding and storing processes of learning and memory information [63]. Although further knowledge is needed, the first studies on the effects of aromatherapy in AD symptoms came along with encouraging results [64].

This has been the base for several studies that evaluate the efficacy as protective agents in AD models of essential oils from Traditional Chinese Medicine herbs. For instance, Liu et al. (2010) demonstrated ameliorative effects on an $A\beta_{1-40}$-induced AD rat model displayed by perfume stimulating olfaction with volatile essential oil of *Acorus Gramineus* (each session lasting for 60 min/day, 42 consecutive days). Dividing rats into groups of ten animals with different treatments, they concluded that a 60 min/day exposure to the volatile oil significantly increase the learning-memory ability (as revealed by Morris Water Maze test), decrease MDA content and increase SOD and glutathione peroxidase activities and weight of brain in AD rats [65]. The mechanism of action for such activity was suggested to be due to a change of a certain neurotransmitter expression in hippocampus, or the neurotransmitters of the olfactory system, influencing change of olfactory conductive signals, so as to strengthen learning and memory functions [65].

Similarly, SuHeXiang Wan is a Chinese traditional medicinal prescription (Storax Pill), consisting of 9 crude herbs, that has revealed neuroprotective effects; its essential oil may have potential as a therapeutic inhalation drug for the prevention and treatment of AD. Actually, in an $A\beta_{1\text{-}40}$-induced *in vivo* model of AD, it turned out to improve memory impairment and normalize biochmical markers in the hippocampus of mice, such as JNK, p38 and Tau phosphorylation. *In vitro*, one hour pretreatment with the essential oil suppressed Aβ-induced apoptosis and ROS production via an up-regulation of hemeoxygenase-1 (HO-1) and Nrf2 expression in SH-SY5Y human neuroblastoma cells [66]. In addition, the Qingxin Kaiqiao Recipe and its volatile oil improved the learning and memory capabilities of AD rats, which might be achieved through a decrease in the expressions of glial fibrillary acidic protein GFAP, Aß, ß-APP, and caspase-3 in the cortex and hippocampus [67].

The Jordanian "Melissa" (*Aloysia citrodora*) is another plant whose essential oil, extracted from leaves, has demonstrated promising effects in this field. Besides acting as a potent iron chelator and free radical (DPPH) scavenger *in vitro*, it demonstrated nicotinic cholinergic activity by inhibiting the binding of [^{3}H] nicotine to rat forebrain membranes. But most interestingly, it showed neuroprotective effects in an Aβ-induced model of AD that used CAD neuroblastoma cell line as substrate; at concentrations of 10 and 1 μg/mL, the essential oil reversed the neurotoxic effects of both amyloid β peptide and hydrogen peroxide, and increased cell viability. The major chemical components detected in the essential oil of *A. citrodora* (limonene, geranial, neral, 1,8-cineole…) are worthy of further investigation as neuroprotective candidates [68].

In this line of research, Cioanca et al. (2014) have recently reported beneficial properties for the essential oil of *Coriandrum sativum* var. *microcarpum* against the AD-related anxiety and depressive behaviors in a rat model. The exposure of $A\beta_{1\text{-}40}$-treated rats to coriander volatile oil significantly augmented their locomotor activity and ameliorated times of reaction and memory in plus-maze and forced swimming test. It was suggested that these actions could be mediated, at least in part, by the antioxidant potential of the essential oil, as it decreased catalase activity and increased glutathione level in the hippocampus of rats [69]. These results confirmed the findings previously achieved by the same team of researchers, who had reported, under the same experimental conditions, that the inhaled coriander volatile oil (1% and 3%, daily, for 21 days) improved spatial memory formation and enhanced cognitive capacities in water maze tests. Those effects seemed to be mediated via antioxidative effects, since biochemical parameters in hippocampus of rats had also ameliorated in this experiment: significant decrease in SOD and LDH activities, increase in GPx specific activity and attenuation of the increase in MDA level; pretreatment with the essential oil even displayed antiapoptotic activity, as revealed by DNA cleavage patterns [70].

One of the most important features in AD is the generation and deposition of neurotoxic ß-amyloid peptide (Aß). The inhibition of BACE-1 (ß-site APP cleaving enzyme 1), a key enzyme in Aß formation, is considered a promising therapeutic alternative for this disease. Videira et al. (2013) screened several essential oils for their inhibitory activity on BACE-1, and found that those from *Lavandula luisieri* exerted effective inhibition of the recombinant enzyme proBACE-1. Moreover, in an *in vitro* cellular AD model (CHO-APPwt cells), it also displayed inhibition of the intracellular enzyme, without significant toxicity, and reduced the formation of Aß. The main inhibitory activity of the essential oil was assigned to the monoterpenic ketone 2,3,4,4-tetramethyl-5-methylene-cyclopent-2-enone, one of the distinctive components of *L. luisieri*, which was also active at inhibiting BACE-1 in other cell lines (e.g., SH-SY5Y cells stably transfected with APPwt) [71]. The entire oil and the referred component were further tested on a triple transgenic mouse model of Alzheimer's disease and, under the stablished experimental conditions (32 mg/kg/day for the essential oil and 92 mg/kg/day for the compound), significant reductions in $A\beta_{40}$ levels were observed in the brains of treated animals compared to control, representing a mean inhibition of approximately 25%. These results imply that tested samples are capable to cross the blood-brain barrier and exert a potential benefit in AD brains [72]. The genus *Lavender* contains other species of much interest regarding their neuroprotective properties. Hritcu et al. (2012) studied the effects of two different lavender essential oils from *Lavandula angustifolia* and *Lavandula hybrida* (Lamiaceae) on neurological capacity in a model of scopolamine-induced dementia rat model. Chronic inhalations of lavender essential oils (60 min daily, for 7 continuous days) significantly reduced anxiety and depressive-like behaviors and also improved spatial memory performances [73].

All in all, the lavender oil is one of the most studied and common essential oils used in therapies. The above-mentioned *in vitro* studies actually give stronger support to the possibilities of lavender oil (especially from *Lavandula angustifolia*) in the management and prevention of Alzheimer´s disease, idea that arose from a clinical trial that was conducted in 2005. That study was developed to evaluate the effects of lavender aromatherapy (hand massage) on cognitive function, emotion and aggressive behavior of elderly with dementia of the Alzheimer's type. It was shown that those patients administrated with the oil for 2 weeks showed ameliorative effects on emotions and aggressive behaviors [74]. Later, Fujii and coworkers (2008) revealed the benefit of therapy with lavender as an intervention for behavioral and psychological symptoms in dementia in patients with AD [75]. A recent open-label trial of daily lavender therapy confirmed the amelioration of such symptoms in 20 patients with frontotemporal lobar degeneration, a disease in which these symptoms are normally more evidenced [76].

On the other hand, melissa (*Melissa officinalis*) is another plant that has demonstrated promising effects in the treatment of cognitive disorders associated to dementia and AD [77]. Its potential was further proved in a placebo-controlled trial that determined the value of aromatherapy with essential oil of *Melissa officinalis* (lemon balm) for agitation in people with severe dementia. In a research with 72 patients over a 4-week period of treatment, treatment with Melissa balm essential oil significantly improved agitation and quality of life indices of patients in comparison to placebo [78].

Zataria multiflora is a member of Lamiaceae family that has been extensively used in Iranian traditional medicine for its beneficial effects on mental abilities. Its essential oil has been evaluated in an $A\beta_{25-35}$-induced AD model in rats, and different doses of 50, 100 and 200 μL/kg (intraperitoneally 30 min before experiments) showed to improve the cognitive capacity of animals without acute toxicity; parameters such as escape latency and traveled distance were measured in the Morris water maze test with better results than in control animals. Thus, it seems that this plant may be a potential valuable source of natural compounds with interest in the therapeutic of AD, and merits further investigations [79].

Moreover, the essential oil obtained by using supercritical CO_2 from alpine needle leaves of *Abies koreana* was investigated for its effects on memory in a scopolamine-induced dementia model in mice. Authors reported that one intraperitoneal injection of doses of 10,

30, 100 and 300 mg/kg, 30 mins before the trial, exerted a memory enhancing effect of 72% in a passive avoidance test, when compared to control. Results suggest that essential oil from *A. koreana* may be an interesting therapeutic agent to be studied in specific AD models. Considering that this essential oil is comprised of a complex mixture of various components (among which the major compounds are elemol, terpinen-4-ol and sabinene), the neuroprotective potential of the pure compounds may also be appealing [80].

Pharmacological studies with the essential oil of lemon [*Citrus limon* (L.) Burms (Rutaceae)] have also exposed its high antioxidant potential and suggested a possible protective effect in oxidative stress-mediated neurodegenerative diseases. By using a mice animal model, Campêlo et al. (2011) demonstrated the capacity of a dose of 0.15 g/kg/day of the essential oil to enhance the antioxidant capacity at hippocampus, which is one of the main areas of the brain affected by oxidative stress in neurodegeneration; it reduced the lipid peroxidation level and nitrite content and increased the GSH levels and the SOD, catalase, and GPx activities in mice hippocampus, with a relative low toxicity [81]. The interest of *Citrus lemon* in the therapeutic of AD has been confirmed by Oboh and coworkers (2014), who demonstrated the potent antioxidant capacities of its essential oil in rat brain homogenates. Besides an exhaustive determination of its components (sabinene, limonene, α-pinene, ß-pinene, neral…), they reported a promising dose-dependent inhibitory activity of acetyl and butyl-cholinesterases and the reduction of pro-oxidant-induced lipid peroxidation. What is more, they proposed the radical scavenging abilities [DPPH, ABTS, NO, $OH^{\cdot}$] and the reducing power as other mechanisms for its antioxidant properties [82]. Besides lemon, other species from the genus *Citrus* could be worth further evaluations on their real potential as therapeutic agents in neurodegenerative disease. Considering the importance of acetylcholinesterase inhibitors for AD treatment, *Citrus angustifolia* essential oil appears as a valuable new flavor with functional properties for foods or nutraceutical products with particular relevance to supplements for the elderly; Tundis et al. (2012) demonstrated an interesting acetylcholinesterase selective inhibition together with a potent radical scavenging capacity in ABTS assay, with more promising results than other *Citrus* species [83]. In this line, due to their implications in AD, there have been several studies measuring the inhibitory activity displayed by essential oils against acetylcholinesterase (AChE) and butylcholinesterase (BChE) enzymes. For example, various species of the genus *Pinus* (*Pinus nigra* subsp. *nigra*, *P. nigra* var. *calabrica*, and *P. heldreichii* subsp. *leucodermis*) were studied for both activities. The outcomes implied that *P. heldreichii* subsp. *leucodermis* exhibited the most promising activities, with with IC_{50} values of 51.1 and 80.6 μg/mL against AChE and BChE, respectively; an interesting activity against AChE was also observed with *P. nigra* subsp. *nigra* essential oil, with an IC_{50} value of 94.4 μg/mL. This study also suggested an interesting activity as acetylcholinesterase inhibitors for one of the main components of *Pinus* sp. essential oils, β-phellandrene, and anti-butylcholinesterase activity for trans-caryophyllene and terpinen-4-ol [84]. The essential oils of several species belonging to the genus *Salvia* (*S. fruticosa*, *S. lavandulaefolia*, *S. officinalis* and *S. officinalis* var. *purpurea*) were also investigated for their inhibitory activities towards acetyl and butyryl-cholinesterases. A dose-dependent inhibition of human cholinesterases (also time-dependent inhibition in the case of human butyrylcholinesterase) was determined using the method of Ellman for the oils of *S. fruticose* and *S. officinalis* var. *purpurea*. The analyses of the chemical composition of the oils and inhibitory activities of their constituents, such as 3-carene and β-pinene, revealed that none of the compounds tested would account for the total activity of the oils and that synergy is likely [85]. Similarly, Loizzo et al. (2009) evaluated the biological activities of *Salvia leriifolia* essential oil which are relevant to the treatment of Alzheimer's disease. They reported that the essential oil (mainly composed of camphor, 1,8-cineole, camphene and alpha-pinene) exhibited interesting cholinesterase activity with IC_{50} values of 0.32 and 0.29 μg/mL for AChE and BChE respectively; in addition, it exerted a promising free radical scavenging activity and inhibition of LPS-induced NO production, with no significant toxicity in a model of monocytic macrophage cell line [86]. The volatile oil of two specimens of *Marlierea racemosa* growing in different areas of Brasil were investigated for their acetylcholinesterase inhibitory activity; they demonstrated a promising effect up to 75% and 35% of inhibition depending on the specimen. The highest activity was correlated with a higher amount in monoterpenes [87]. Apart from anticholinesterase activity, other activities relevant to Alzheimer's disease include antioxidant effects. With this respect, Porres-Martínez and coworkers conducted further research on essential oils from *Salvia lavandulifolia* Valh., which also deserves to be highlighted in this chapter; even though specific AD models were not employed, their work demonstrated promising properties by the essential oil against oxidative stress in cellular models of central nervous system. Two essential oil samples, obtained in different plant density conditions, exhibited a potent antioxidant capacity to scavenge intracellular ROS, to enhance the endogenous antioxidant system (phase-II antioxidant enzymes and glutathione system) and to inhibit lipid peroxidation in a H_2O_2-induced oxidative stress model in astrocytes [88]. Analogous cytoprotective results were evidenced in a cellular model of neurons, with involvement of the activation of Nrf2 transcription factor as a crucial cellular pathway [89]. Besides analyzing the influence of the phenological state of the plant on the chemical composition and on the antioxidant and cytoprotective properties displayed by the essential oils [90], these authors also demonstrated the effects of two of the majors monoterpenes (1,8-cineole and α-pinene) identified in *S. lavandulifolia* essential oil as regulators of cellular redox balance in astrocytes [91]. Then, this species appears to have potential value in neurodegenerative diseases such as AD.

There are more other studies that specifically dealt with the isolated components of some essential oils and their bioactive effects with interest in AD therapy. For instance, Jukic et al. (2007) examined the *in vitro* inhibitory activity exerted by the main constituents of the essential oil obtained from the aromatic plant *Thymus vulgaris* L. on acetylcholinesterase. Among the selected compounds, thymohydroquinone exhibited the strongest AChE inhibitory effect over the range of concentrations, and the activity decreased in the following order: thymohydroquinone (EC_{50} = 0.04 mg/mL) > carvacrol (0.063 mg/mL) > thymoquinone (0.14 mg/mL) > thymol (0.74 mg/mL) > linalool. It is interesting that the AChE inhibitory effect exerted by carvacrol was 10 times stronger than that exerted by its isomer thymol, although thymol and carvacrol have a very similar structure (see Figure 2) [92]. The efficacy of monoterpenes as inhibitors of AChE was later confirmed in an *in vitro* study including selected components of essential oils, which carry a variety of important functional groups, through an adapted version of Ellman's colorimetric assay. It was evidenced that 1,8-cineole, carvacrol, myrtenal and verbenone apparently inhibited AChE; the highest inhibitory activity was observed for myrtenal (IC_{50} = 0.17 mm) [93].

Table 1. The most outstanding beneficial effects of essential oils on neurodegeneration and Alzheimer´s disease models

Essential oil	Main constituents	Model of study	Administration	Effects	Reference
Acorus gramineus essential oil	Not specified	$Aß_{1-40}$–induced AD rat model	Perfume smoking sessions with the essential oil: 60 min/day for 42 consecutive days	↑ Learning memory ability in Morris Water Maze test ↓ MDA content and ↑ SOD and GPx activities in rat brain	[65]
Su He Xiang Wan essential oil	9 crude herbs from Traditional Chinese Medicine	$Aß_{1-42}$–induced AD mice model	Inhalation for 6 h of 2 g of fragrance oil per day, 14 consecutive days	Improvement of memory impairment (Y-maze test) ↓ JNK, p38 and Tau-phosphorylation in hippocampus	[66]
		$Aß_{1-42}$–induced AD neurotoxicity in SH-SY5Y human neuroblastoma cells	Pretreatment of cells for 1 hour with the essential oil before Aß exposure	↓ Aß-induced apoptosis and ROS production ↑ Up-regulation of hemeoxygenase-1 and Nrf2 expression	
Aloysia citrodora essential oil (from fresh and dried leafs)	Limonene, geranial, neral, 1,8-cineol, curcumene, spathulenol, caryophyllene oxide.	H_2O_2 and Aß-induced neurotoxicity in CAD neuroblastoma cell line	Pretreatment of cells for 24 h with 1 and 10 µg/mL prior to neurotoxic insults	↑ Cell survival ↓ [^{3}H]-nicotine binding to rat forebrain membranes ↑ Iron chelation and antioxidant activity *in vitro* (DPPH radical scavenging capacity)	[68]
Coriander sativum var. *microcarpum* essential oil	Linalool, α-pinene, γ-terpinene, geranyl acetate, camphor and geraniol	$Aß_{1-42}$–induced AD rat model	Exposure to coriander volatile oil 1 and 3% for 60 min/day during 21 days	Anti-depressant and anxiolytic effects in rats Improvement in memory and behavioral impairments ↓ catalase activity and ↑ glutathione levels in hippocampus	[69]
				Improved spatial memory formation and enhanced cognitive capacities ↑ GPx activity and ↓ SOD and LDH activities and MDA levels Antiapoptotic effect: changes in DNA cleavage patterns	[70]

Table 1. (Continued)

Essential oil	Main constituents	Model of study	Administration	Effects	Reference
Lavandula luisieri essential oil	Monoterpenic ketone 2,3,4,4-tetramethyl-5-methylene-cyclopent-2-enone, and others	3xTg-AD mouse model	32 mg/kg/day of the essential oil	No significant toxicity on cells in culture nor *in vivo* Inhibitory activity towards BACE-1 *in vitro* and *in vivo*, with reduced production and levels of Aß$_{40}$	[72, 73]
		Chinese hamster ovary (CHO) cells transfected with APPwt	6 hour treatment with the essential oil at 90 μg/mL		
Lavandula angustifolia and *Lavandula hybrida* essential oils	Linalool, linalyl acetate, cineole, terpinen-4-ol and camphor	Scopolamine-induced dementia rat model	Chronic exposures (inhalation of essential oil vapors) for 60 min/day, 7 consecutive days	↓ Anxiety and depression-like behaviors ↑ Spatial memory performances	[74]
Melissa officinalis essential oil	Not specified	72 patients suffering from AD-like dementia	4 week-period treatment Vs placebo	Significant amelioration in agitation and quality of life of patients	[77]
Zataria multiflora essential oil	Carvacrol, thymol, p-cymene, linalool, γ-terpinene	Aß$_{25\text{-}35}$–induced AD rat model	Intraperitoneal injection (30 min prior to Aß exposure) of doses 50, 100 and 200 μL/kg	Amelioration of cognitive deficits and motor signs in Morris Water Maze test	[79]
Abies koreana essential oil (supercritical CO_2 fluid extract)	Elemol, terpinen-4-ol, sabinene, cadienol, α-terpinenol, α-pinene, γ-terpinene	Scopolamine-induced dementia mouse model	Intraperitoneal injection of essential oil in a dose of 100 mg/kg	Memory enhancing activity	[80]
Citrus lemon essential oil	Sabinene, limonene, α-pinene, ß-pinene, neral, etc.	Swiss mice	Intraperitoneal injection of doses 0.05, 0.1 and 0.15 g/kg	↓ Lipid peroxidation level and nitrite content ↑ Glutathione levels and SOD, CAT and GPx activities in mice hippocampus	[81]
		Rat brain homogenates	-	Inhibitory activity of AChE and BChE	[82]

4. Essential Oils as Neuroprotective Agents in Parkinson's Disease Models

Parkinson´s disease (PD) is a chronic, multifactorial and progressive neurologic condition that affects around six million people worldwide, normally over 60 years of age, thus being by far the second most common degenerative disorder affecting the central nervous system (after Alzheimer's disease) and the most common movement disorder, with geographic and genetic variations [94]. PD is typically diagnosed by a clinical triad: tremor, akinesia and rigidity, and both the clinical and pathological features are associated with the loss of neurons in certain brain areas, precisely with the neurodegeneration of dopaminergic neurons in the substantia nigra; actually, when PD is diagnosed by the onset of symptoms, neuronal loss is estimated to be 60 to 70 percent. The reduction in dopamine levels (in the substantia nigra and striatum) when dopaminergic neurons decline leads to the characteristic motor disturbances of the disease, including gait disturbances, postural bradykinesia, instability, rigidity and tremor, and non-motor symptoms such as sleep disorders, constipation, depression and genitourinary problems [95]. Besides neuronal death, a pathological hallmark of PD lies in the presence of the Lewy bodies (eosinophilic intracytoplasmic proteinaceous inclusions), and dystrophic neurites (Lewy neurites) presented in the surviving neurons, that are comprised of depositions of anomalous α-synuclein protein filaments and likewise contribute to the pathogenesis of PD [96].

As for AD, oxidative stress is believed to play an important role in neuronal degeneration and nigral cell death in PD by compromising the integrity of vulnerable neurons, as revealed by an increase in lipid peroxidation and a widespread increase in protein and DNA oxidation in brains of PD patients. Plausible causes that underlie the excessive oxidative stress include increased dopamine metabolism (that can yield abundant hydrogen peroxide, hydroxyl and superoxide radicals and other ROS), high reactive iron levels and impaired antioxidant defense system (i.e., decreased levels of reduced glutathione). What is more, consistent findings highlight mitochondrial dysfunction as a central factor in the pathogenesis of sporadic PD, and in particular, impaired mitochondrial complex-I of the respiratory chain in the subtantia nigra [97]; with this respect, free radical scavengers and antioxidants targeted to mitochondria have attracted attention as agents capable to prevent disease progression. Among other investigated drugs, natural products are revealing promising effects in PD models [98, 99].

Concerning the relevance of plant essential oils as potential neuroprotective agents in PD, little research has been conducted, but some components of the essential oils should merit further research.

For instance, (±) isoborneol is a monoterpenoid alcohol present in the essential oils of numerous medicinal plants (including valerian, chamomile and lavender) and is a known antioxidant, which has been investigated for its neuroprotective effect against 6-OHDA-induced cell death in human neuroblastoma SH-SY5Y cells. In such *in vitro* model of PD, cell pretreatment with isoborneol significantly prevented cells from death and diminished the induced overproduction of ROS and the increase in intracellular calcium; moreover, isoborneol pretreatment reversed the enhanced apoptosis via reductions in Bax/Bcl-2 ratio, caspase-3 activity and mitochondrial release of cytochrome C into the cytosol. These results, taken together with the capacity of isoborneol to downregulate the activity of c-Jun Nterminal kinase and to induce the activation of protein kinase C (PKC), suggest a protective function of isoborneol in PD which is dependent upon its antioxidant potential [100].

Similarly, zingerone and eugenol, the main components of the essential oils extracted from ginger and cloves, were examined regarding their protective effects on a 6-OHDA-induced mouse model of PD. Eugenol administration 3 days before and 7 more days following one intracerebroventricular 6-OHDA injection prevented the reduction of striatal dopamine and its metabolites, reduced the elevated lipid peroxidation and enhanced the antioxidant systems (glutathione and L-ascorbate) in the striatum of mice [101]. In the same way, zingerone administration 1 h before and for 6 more days following 6-OHDA injection elevated the striatal dopamine levels and increased serum $O_2^{\cdot -}$ scavenging activity, proposing an increased systemic superoxide dismutase activity [102]. Effects of the pretreatment with these two compounds in this model suggest their possible value in the therapy of Parkinson's disease. However, deeper studies are needed, since opposite results were found for a post-treatment with these compounds in the same model (administrations only after the intracerebroventricular 6-OHDA injection), where they worsened the neurotoxic effects [103].

Conclusion and Future Prospects

With an older population worldwide, the number of people suffering from neurodegenerative diseases such as Alzheimer´s and Parkinson´s diseases is exponentially increasing. Since these conditions are lacking effective treatments, the need for the search of new drugs that help to cure or avoid and delay the onset of these diseases is one of the main challenges of the current century.

The investigations on the neuroprotective effect of essential oils against the oxidative stress in PD and AD have demonstrated their promising and potential efficacy in preliminary studies including *in vitro* and *in vivo* models of research. However, it is still mandatory the realization of deeper *in vivo* studies as well as clinical trial to really confirm the therapeutic efficacy of essential oils in this two common age-related neurodegenerative illnesses.

What is more, most of the studies referred in this chapter have been conducted with the total essential oil or with the major compound present in it, without evaluating the effects of other compounds in lower concentrations. Considering the complex mixtures of molecules that the essential oils represent, this fact may be even more important, as the effect of the total essential oil could be considered as the global effect of all its components, including the synergistic effects among the individual compounds.

In conclusion, essential oils appear as a promising therapeutic strategy for slowing or reversing AD and PD, but more and exhaustive studies are required in order to understand the real potential in neuroprotective therapy.

REFERENCES

[1] M. Czerska, K. Mikołajewska, M. Zieliński, J. Gromadzińska, W. Wąsowicz. *Med. Pr.* 66, 3 (2015).
[2] R.K. Gupta, A.K. Patel, N. Shah, A.K. Chaudhary, U.K. J ha, U.C. Yadav, P.K. Gupta, U. Pakuwal. *Asian Pac. J. Cancer Prev.* 15, 11 (2014).
[3] E. Niki. *Biofactors.* 34, 2 (2008).
[4] J. Wong-ekkabut, Z. Xu, W. Triampo, I-M. Tang, D.P. Tieleman, L. Monticelli. *Biophys J.* 93, 12 (2007).
[5] A. Ayala, M.F. Muñoz, S. Argüelles. *Oxid. Med. Cell Longev.* 2014 (2014).
[6] T.C. Squier. *Exp. Gerontol.* 36, 9 (2001).
[7] R.A. Dunlop, U.T. Brunk, K.J. Rodgers. *IUBMB Life.* 61, 5 (2009).
[8] P. Sharma, A.B. Jha, R.S. Dubey, M. Pessarakli. *J. Bot.* 2012 (2012).
[9] A. Barzilai, K. Yamamoto. *DNA* 3, 8-9 (2004).
[10] T.B. Kryston, A.B. Georgiev, P. Pissis, A.G. Georgakilas. *Mutat. Res.* 711, 1-2 (2011).
[11] Q. Kong, C.L. Lin. *Cell Mol. Life Sci.* 67, 11 (2010).
[12] Z. Li, J. Wu, C.J. Deleo. *IUBMB Life.* 58, 10 (2006).
[13] V.I. Lushchak. *Chem. Biol. Interact.* 224C (2014).
[14] S.I. Liochev. *Free Radic. Biol. Med.* 60 (2013).
[15] A. Phaniendra, D.B. Jestadi, L. Periyasamy. *Indian J. Clin. Biochem.* 30, 1 (2015).
[16] V.G. Grivennikova, A.D. Vinogradov. *Biochemistry (Mosc).* 78, 13 (2013).
[17] K. Brieger, S. Schiavone, F.J. Jr Miller, K.H. Krause. *Swiss Med. Wkly.* 142, w13659 (2012).
[18] C.N. Paiva, M.T. Bozza. *Antioxid. Redox Signal.* 20, 6 (2014).
[19] D. Vara, G. Pula. *Curr. Mol Med.* 14, 9 (2014).
[20] B. Uttara, A.V. Singh, P. Zamboni, R.T. Mahajan. Curr. Neuropharmacol. 7, 1 (2009).
[21] N.R. Madamanchi, A. Vendrov, M.S. Runge. *Arterioscler. Thromb. Vasc. Biol.* 25, 1 (2005).
[22] V. Sosa, T. Moliné, R. Somoza, R. Paciucci, H. Kondoh, M.E. LLeonart. *Ageing Res. Rev.*12, 1 (2013).
[23] B. Halliwell. *Nutr. Rev.* 70, 5 (2012).
[24] M. Wojcik, I. Burzynska-Pedziwiatr, L. A. Wozniak. *Curr. Med. Chem.* 17, 28 (2010).
[25] E. Niki. *Biomed. J.* 37, 3 (2014).
[26] D. P. Gelain, R. J. Dalmolin, V. L. Belau, J. C. Moreira, F. Klamt, M. A. Castro. *Front Biosci (Landmark Ed).* 14 (2009).
[27] K. Aoyama, T. Nakaki. *Molecules.* 20, 5 (2015).
[28] F. Gu, V. Chauhan, A. *Chauhan. Curr. Opin. Clin. Nutr. Metab. Care.* 18, 1 (2015).
[29] D.M. Kasote, S.S. Katyare, M.V. Hegde, H. Bae. *Int. J. Biol. Sci.* 11, 8 (2015).
[30] I. Sadowska-Bartosz, G. Bartosz. *Biomed. Res. Int.* 2014 (2014).
[31] P. Tongnuanchan, S. Benjakul. *J. Food Sci.* 79, 7 (2014).
[32] C. Zitzelsberger, G. Buchbauer. *Nat. Prod. Commun.* 10, 6 (2015).
[33] D. Grulova, L. De Martino, E. Mancini, I. Salamon, V. De Feo. *J. Sci. Food Agric.* 95, 3 (2015).
[34] K. El Hamrouni-Aschi, M.L. Khouja, M. Boussaid, N. Akrimi, L. Toumi. *Chem. Biodivers.* 10, 6 (2013).

[35] X.H. Zhang, J.A. da Silva, Y.X. Jia, J.T. Zhao, G.H. Ma. *Nat. Prod. Commun.* 7, 1 (2012).
[36] M.G. Motto, N.J. Secord. *J. Agric. Food Chem.* 33, 5 (1985).
[37] J. Bernath. Aromatic plants. In Cultivated Plants, Primarily as Food Sources. Gyorgy Fuleky (Editor). Volume 2. 2009.
[38] [R.S. Verma, R.K. Verma, A. Chauhan, A.K. Yadav. *Indian J. Pharm. Sci.* 73, 2 (2011).
[39] D. Kumar, M. Sukapaka, G.D. Babu, Y. Padwad. *PLoS One.* 10, 8 (2015).
[40] D.C. Cabrera, G.L. Gomes, A. Flach, L.A. da Costa, G.R. Rosa, N.F. de Moura. *Nat. Prod. Res.* 29, 7 (2015).
[41] H.S. Choi, M. Sawamura. *J. Agric. Food Chem.* 48, 10 (2000).
[42] A. El Asbahani, K. Miladi, W. Badri, M. Sala, E.H. Aït Addi, H. Casabianca, A. El Mousadik, D. Hartmann, A. Jilale, F.N. Renaud, A. Elaissari. *Int. J. Pharm.* 483, 1-2 (2015).
[43] J. Vergis, P. Gokulakrishnan, R.K. Agarwal, A. Kumar. *Crit. Rev. Food Sci. Nutr.* 55, 10 (2015).
[44] A.D. Melo, A.F. Amaral, G. Schaefer, F.B. Luciano, C. de Andrade, L.B. Costa, M.H. Rostagno. *Can. J. Vet. Res.* 79, 4 (2015).
[45] D.H. Gilling, M. Kitajima, J.R. Torrey, K.R. Bright. *J. Appl. Microbiol.* 116, 5 (2014).
[46] M. Bogavac, M. Karaman, Lj. Janjušević, J. Sudji, B. Radovanović, Z. Novaković, J. Simeunović, B. Božin. *J. Appl. Microbiol.* 119, 3 (2015).
[47] E. Khodarahmi, G.H. Asghari, F. Hassanzadeh, M. Mirian, G.A. Khodarahmi. *Res. Pharm Sci.* 10, 2 (2015).
[48] K. Fan, X. Li, Y. Cao, H. Qi, L. Li, Q. Zhang, H. Sun. *Anticancer Drugs.* 26, 8 (2015).
[49] M. Melušová, S. Jantová, E. Horváthová. *Interdiscip. Toxicol.* 7, 4 (2014).
[50] A. Russo, V. Cardile, A.C. Graziano, C. Formisano, D. Rigano, M. Canzoneri, M. Bruno, F. Senatore. *Nat. Prod. Res.* 29, 17 (2015).
[51] A.M. Saleh, A. Aljada, S.A. Rizvi, A. Nasr, A.S. Alaskar, J.D. Williams. *BMC Complement. Altern. Med.* 14 (2014).
[52] J.D. Cha, J.Y. Kim. *Molecules.* 17, 4 (2012).
[53] J.O. Akolade, L.A. Usman, O.E. Okereke, N.O. Muhammad. J. Med. Food. 17, 10 (2014).
[54] H.E. Joo, H.J. Lee, E.J. Sohn, M.H. Lee, H.S. Ko, S.J. Jeong, H.J. Lee, S.H. Kim. *Biosci. Biotechnol. Biochem.* 77, 10 (2013).
[55] S.C. Lee, W.X. Xu, L.Y. Lin, J.J. Yang, C.T. Liu. *J. Agric. Food Chem.* 61, 10 (2013).
[56] M. Höferl, I. Stoilova, J. Wanner, E. Schmidt, L. Jirovetz, D. Trifonova, V. Stanchev, A. Krastanov. *Nat. Prod Commun.* 10, 6 (2015).
[57] M.R. Castro, F.N. Victoria, D.H. Oliveira, R.G. Jacob, L. Savegnago, D. Alves. *Pharm. Biol.* 53, 2 (2015).
[58] M. Goedert, M. G. Spillantini. *Science* 3 (2006).
[59] J.L. Cummings. *N. Engl. J. Med.* 351 (2004).
[60] M. Pohanka. *Curr. Med. Chem.* 21, 3 (2014).
[61] A.M. Swomley, D. A. Butterfield. *Arch. Toxicol.* 89, 10 (2015).
[62] Y. Zhao, B. Zhao. *Front. Biosci. (Elite Ed).* 4 (2012).
[63] M.T. Hsieh, W.H. Peng, C.R. Wu, W.H. Wang. *Phytother. Res.* 14 (2000).

[64] D. Jimbo, Y. Kimura, M. Taniguchi, M. Inoue, K. Urakami. *Psychogeriatrics*. 9, 4 (2009).

[65] Z.B. Liu, W.M. Niu, X.H. Yang, Y. Wang, W.G. Wang. *J. Tradit. Chin. Med.* 30, 4 (2010).

[66] S. Jeon, J. Hur, H.J. Jeong, B.S. Koo, S.C. Pak. *Am. J. Chin. Med.* 39, 5 (2011).

[67] H.Y. Hu, L. Lei, X. Chen, W.H. Wang, X.Y. Zhang. *Zhongguo Zhong Xi Yi Jie He Za Zhi*. 33, 7 (2013).

[68] S. Abuhamdah, R. Abuhamdah, M. J. Howes, S. Al-Olimat, A. Ennaceur, P. L. Chazot. *J. Pharm. Pharmacol.* 67, 9 (2015).

[69] O. Cioanca, L. Hritcu, M. Mihasan, A. Trifan, M. Hancianu. *Physiol. Behav.* 131 (2014).

[70] O. Cioanca, L. Hritcu, M. Mihasan, M. Hancianu. *Physiol Behav.* 120 (2013).

[71] S.Y. Lee. *Taehan Kanho Hakhoe Chi.* 35, 2 (2005).

[72] R. Videira, P. Castanheira, M. Grãos, L. Salgueiro, C. Faro, C. Cavaleiro. *Flavour Frag. J.* 28, 6 (2013).

[73] R. Videira, P. Castanheira, M. Grãos, R. Resende, L. Salgueiro, C. Faro, C. Cavaleiro. *J. Nat. Prod.* 77 (2014).

[74] L. Hritcu, O. Cioanca, M. Hancianu. *Phytomedicine* 19 (2012).

[75] M. Fujii, R. Hatakeyama, Y. Fukuoka, T. Yamamoto, R. Sasaki, M. Moriya, M. Kanno, H. Sasaki. *Geriatr. Gerontol. Int.* 8, 2 (2008).

[76] T. Kimura, J. Takamatsu. *Geriatr. Gerontol. Int.* 13, 2 (2013).

[77] S. Akhondzadeh, M. Noroozian, M. Mohammadi, S. Ohadinia, A.H. Jamshidi, M. Khani. *J. Neurol. Neurosurg. Psychiatry*. 74, 7 (2003).

[78] C.G. Ballard, J.T. O'Brien, K. Reichelt, E.K. Perry. *J. Clin. Psychiatry*. 63, 7 (2002).

[79] N. Majlessi, S. Choopani, M. Kamalinejad, Z. Azizi. *CNS Neurosci. Ther*. 18, 4 (2012).

[80] K. Kim, Y. Bu, S. Jeong, J. Lim, Y. Kwon, D.S. Cha, J. Kim, S. Jeon, J. Eun, H. Jeon. *Biosci. Biotechnol. Biochem*. 70, 8 (2006).

[81] L.M. Campêlo, F.C. Gonçalves, C.M. Feitosa, R.M. de Freitas. *Pharm. Biol.* 49, 7 (2011).

[82] G. Oboh, T.A. Olasehinde, A.O. Ademosun. *J. Oleo Sci.* 63, 4 (2014).

[83] R. Tundis, M.R. Loizzo, M. Bonesi, F. Menichini, V. Mastellone, C. Colica, F. Menichini. *J. Food Sci.* 77, 1 (2012).

[84] M. Bonesi, F. Menichini, R. Tundis, M.R. Loizzo, F. Conforti, N.G. Passalacqua, G.A. Statti, F. Menichini. *J. Enzyme Inhib. Med. Chem.* 25, 5 (2010).

[85] S.U. Savelev, E.J. Okello, E.K. Perry. *Phytother. Res.* 18, 4 (2004).

[86] M.R. Loizzo, F. Menichini, R. Tundis, M. Bonesi, F. Conforti, F. Nadjafi, G.A. Statti, N.G. Frega, F. Menichini. *J. Oleo Sci.* 58, 8 (2009).

[87] A. Souza, M.C. Silva, E.M. Cardoso-Lopes, I. Cordeiro, M.E. Sobral, M.C. Young, P.R. Moreno. *Nat. Prod. Commun.* 4, 8 (2009).

[88] M. Porres-Martínez, E. González-Burgos, M.E. Carretero, M.P. Gómez-Serranillos. *Food Res. Int.* 54, 1 (2013).

[89] M. Porres-Martínez, E. González-Burgos, M.E. Carretero, M.P. Gómez-Serranillos. *Food Chem. Toxicol.* 80 (2015).

[90] M. Porres-Martínez, E. González-Burgos, M.E. Carretero, M.P. Gómez-Serranillos. *Ind. Crops Prod.* 53 (2014).

[91] M. Porres-Martínez, E. González-Burgos, M.E. Carretero, M.P. Gómez-Serranillos. *Pharm. Biol.* 53, 6 (2015).

[92] M. Jukic, O. Politeo, M. Maksimovic, M. Milos, M. Milos. *Phytother. Res.* 21, 3 (2007).

[93] D. Kaufmann, A.K. Dogra, M. Wink. *J. Pharm. Pharmacol.* 63, 10 (2011).

[94] A. Wright Willis, B.A. Evanoff, M. Lian, S.R. Criswell, B.A. Racette. *Neuroepidemiology*. 34, 3 (2010).

[95] R. Xia, Z.H. Mao. *Neurosci. Bull.* 28 (2012).

[96] L.V. Kalia, S.K. Kalia, P.J. McLean, A.M. Lozano, A.E. Lang. *Ann. Neurol.* 73 (2013).

[97] P. Jenner. *Ann. Neurol.* 53 (2003).

[98] E. González-Burgos, C. Fernández-Moriano, M.P. Gómez-Serranillos. *J. Neuroimmune Pharmacol.* 10, 1 (2015).

[99] C. Fernández-Moriano, E. González-Burgos, M.P. Gómez-Serranillos. *Oxid. Med. Cell. Longev.* 2015 (2015).

[100] L.L. Tian, Z. Zhou, Q. Zhang, Y.N. Sun, C.R. Li, C.H. Cheng, Z.Y. Zhong, S.Q. Wang. *Cell. Physiol. Biochem.* 20, 6 (2007).

[101] H. Kabuto, M. Tada, M. Kohno. *Biol. Pharm. Bull.* 30 (2007).

[102] H. Kabuto, M. Nishizawa, M. Tada, C. Higashio, T. Shishibori, M. Kohno. *Neurochem. Res.* 30 (2005).

[103] H. Kabuto, T.T. Yamanushi. *Neurochem. Res.* 36, 12 (2011).

In: Essential Oils
Editor: Miranda Peters
ISBN: 978-1-63484-351-5

Chapter 7

ESSENTIAL OILS OF SOME ALGERIAN ENDEMIC AND MEDICINAL PLANTS: VARIABILITY OF THEIR CHEMICAL COMPOSITIONS AND THEIR POTENT AS NATURAL ANTIOXIDANTS

Nadhir Gourine*
Laboratory of Research of Fundamental Sciences,
University Amar Télidji of Laghouat, Laghouat, Algeria

ABSTRACT

Essential oils (EOs) have been long recognized for their medicinal uses: antibacterial, antifungal, antiviral, insecticidal and their antioxidant properties. The increased interest in alternative natural substances is driving the research community to find new uses and applications of these substances. Algeria, rich in indigenous herbal resources which grow on its varied topography and under changing climatic conditions permitting the growth of almost 3000 plant species (Cheriti et al. 2006). In the semi-arid regions, local traditional pharmacopoeia continues to be an important source of remedies for primary healthcare, so the region affords ample scope for studies concerning various aspects of folk medicine (Cheriti et al. 2006). The purpose of this chapter is to provide an overview on the variability of the chemical composition of the essential oil of some widely common used medicinal and endemic plants growing in Algeria: *Pistacia atlantica, Pituranthos scoparius* and *Rhanterium adpressum*. In the other hand, the growing interest in the substitution of synthetic food antioxidants by natural ones has fostered research on vegetable sources and the screening of raw materials for identifying new antioxidants. In this context, a survey of the antioxidant activities of theses oils is also reported. Furthermore, and due to the lack of a standard protocol of measuring antioxidant activity *via* DPPH assay, which is by the way the most employed *in vitro* technique, it is difficult to compare the results reported from different research groups; in this context, a new approach for effective comparison and expression is proposed.

* E-mail: gourine.nadir@gmail.com or n.gourine@mail.lagh-univ.dz.

INTRODUCTION

One of the most important characteristics of spontaneous plants in Algeria is they offer a very large variability of the chemical composition of their essential oils. In this context, a selection of some wide common plants growing in arid and semi arid region of Algeria is presented in this chapter. The selected plants are *Pistacia atlantica, Pituranthos scoparius* and *Rhanterium adpressum*. Furthermore, the antioxidant activities of their essential oils are discussed.

Pistacia Atlantica

The genus *Pistacia* (Anacardiaceae) is widely distributed in the Mediterranean area (Bailey 1977). *Pistacia atlantica* Desf. is a tree located in North Africa, which can reach over 15 m in height and grows in arid and semi-arid areas (Abdelmalek 2000). The trees are also known in Algeria in the local Arabic language as "Butom"; the fruits called by the locals as "*El-Khodiri*" are edible and sold in markets. *Pistacia Atlantica* is valued because it is the source of mastic gum, an exudate which strengthens gums, deodorises breath, fights coughs, chills and stomach diseases (Bellakhdar 1997). Moreover, the galls of *Pistacia atlantica* are used as an embalming agent by rural inhabitants. They are also known in Arabic as ''Afse'' and are sold in local traditional markets.

Pistacia atlantica is widely spread over different regions in south Algeria (arid and semi arid regions); the main regions are: Djelfa "Senalba, Aïn Oussera, Messaad", Laghouat "south" and Ghardaïa "in Oued M'zab" (Monjauze 1980, Seigne 1985).

Only a few reports were carried out on the chemical composition of essential oils of *Pistacia atlantica* (Table 1).

Concerning Algeria (North Africa in a Mediterranean zone), the first report on this tree dealt with the chemical composition of leaves, fruits and galls.

Figure 1. Tree of *Pistacia atlantica* Desf. from the region of Aïn-Ousserra (Djelfa); Photo taken in October 2007.

Table 1. Variability of the chemical composition of essential oils of *Pistacia atlantica*

Tree studied part	Origin of country	Main components	Ref.
Leaves	Morocco	Terpinen-4-ol (21.7%) Elemol (20.0%)	(Barrero et al. 2005)
	Greece "male tree"	Germacrene D (18.2%) Terpinen-4-ol (17.3%) (*E*)-Caryophyllene (17.3%) *γ*-Terpinene (10.5%) Sabinene (7.5%) *α*-Terpinene (5.9%) *α*-Pinene (5.1%)	(Tzakou, Bazos, and Yannitsaros 2007)
		p-Mentha-1(7),8-diene (41.1%) Germacrene D (19.9%) *δ*-Cadinene (5.7%)	
	Greece "female tree"	Myrcene (17.8%) Germacrene D (15.7%) Terpinen-4-ol (11.6%) (*E*)-Caryophyllene (11.2%) *γ*-Terpinene (8.7%) Sabinene (7.8%) *α*-Terpinene (5.1%)	
		Myrcene (24.8%) Germacrene D (17.8%) (*E*)-Caryophyllene (9.0%) Terpinen-4-ol (6.0%) Sabinene (5.2%) *δ*-Cadinene (4.8%)	
	Algeria	*α*-Pinene (32.6–54.7%) *β*-Pinene (9.6–20.2%)	(Mecherara-Idjeri et al. 2008)
		Chemotype 1: α-pinene/camphene (10.9-66.6%/3.8-20.9%) Chemotype 2: *α*-terpinen-4-ol (10.8-16.0%) Chemotype 3: *δ*-3-carene (16.4-56.2%)	(Gourine, Bombarda, et al. 2010)
Galls	Algeria	*α*-Pinene (48.3%) *β*-Pinene (8.0%)	(Mecherara-Idjeri et al. 2008)
		Chemotype 1: *α*-pinene (59.01%), *β*-pinene (13.26%) Chemotype 2: Δ^3-carene (75.34%)	(Gourine, Sifi, et al. 2011)

Table 1. (Continued)

Tree studied part	Origin of country	Main components	Ref.
		Chemotype 1: α-pinene (12.5-81.5%) Chemotype 2: sabinene (8.1-28.0%); terpinen-4-ol (8.7–20.9%); α-pinene (14.2-32.8%)	(Sifi et al. 2015)
Fruits	Algeria	α-Pinene (50.6%) β-Pinene (13.5%)	(Mecherara-Idjeri et al. 2008)
	Morocco	Bornyl acetates (21.5%) α-Pinene (3.8%)	(Barrero et al. 2005)
	Greece unripe (green) "female tree"	Terpinen-4-ol (25.7%) Myrcene (20.2%) Sabinene (14.9%) *p*-Cymene (10.0%) α-Pinene (4.5%)	(Tzakou, Bazos, and Yannitsaros 2007)
		Myrcene (34.5%) Sabinene (19.5%) Terpinen-4-ol (8.9%) Germacrene D (6.8%) α-Pinene (6.6%) γ-Terpinene (5.0%)	
Leaf-buds	Greece "male tree"	Sabinene (52.1%) α-Pinene (11.6%) Terpinen-4-ol (10.1%) γ-Terpinene (4.6%)	
	Greece "male tree"	*p*-Mentha-1(7),8-diene (42.4%) Germacrene D (20.3%) β-Cubebene (5.1%) α-Ylangene (4.3%)	
Resin	Morocco	α-Pinene (42.9%) β-Pinene (13.2%)	(Barrero et al. 2005)

This report revealed the presence of a unique chemotype which was rich in α-pinene and β-pinene for the investigated parts (leaves, fruits and galls); shortly afterwards, some other more detailed reports which involved an investigation at a large scale, i.e., it used large numbers of samples extended to different locations in Algeria, revealed the presence of different chemotypes for the leaves and for the galls.

Figure 2. Leaves of *Pistacia atlantica* Desf. (Front and back positions).

Figure 3. Galls of *Pistacia atlantica* Desf.

Figure 4. Early stage formation of Galls of *Pistacia atlantica* Desf. tree from the region of Aïn-Ousserra (Djelfa); Photo taken in Augest 2010.

For the leave essential oils three chemotypes were identified using multivariable statistical analysis (Principal component analysis PCA), and they were α-pinene/camphene chemotype (10.9-66.6%/3.8-20.9%), α-Terpinen-4-ol chemotype (10.8-16.0%) and δ-3-carene chemotype (16.4-56.2%), respectively.

For the galls essential oils three chemotypes were also identified, and they were α-pinene chemotype (12.5-81.5%), α-pinene/terpinen-4-ol/sabinene chemotype (14.2-32.8%/8.7-20.9%/8.1-28.0%) and δ-3-carene chemotype (75.34%), respectively. The comparison of the major identified components of essential oils of leaves, and galls in Algeria with the nearest country like Morocco, Greece, shows very different compositions profiles; moreover, in Algeria and unlike some other countries, different chemotypes were identified, which expresses a large variability of the composition of these essential oils. Furthermore, the δ-3-carene was a unique rare chemotype identified only in Algeria.

The antioxidant activities of the essential oil of *Pistacia atlantica* leaves essential oils were studied for different regions using a big number of samples (Gourine, Bombarda, et al. 2010). The seasonal variation of these activities was also investigated for the region of Laghouat from June to October (Gourine, Yousfi, Bombarda, Nadjemi, and Gaydou 2010). Two methods of evaluation antioxidant activities were employed.

The first method was DPPH (*2,2- di-phenyl-1-picrylhydrazyl*) free radical-scavenging assay, this method is based on the reduction of alcoholic DPPH solutions in the presence of a hydrogen donating antioxidant. The second method was the FRAP (Ferric Reducing Antioxidant Power) which is method based on the determination of ferric reducing antioxidant power FRAP.

Figure 5. Ripe and unripe fruits of *Pistacia atlantica* Desf.

The two assays employed (DPPH and FRAP) are complementary *in vitro* tests of antioxidant activity. i.e., the DPPH test measures the power of scavenging free radicals in organic system environment, whereas the FRAP test measures the power of reducing of free metalic ions in a water system environment. This last method was initially developed to assay plasma antioxidant capacity, then it was extended to plant extracts too. When DPPH assay employed for the investigation of the antioxidant activity of the essential oil samples belonging to different region from Algeria (Laghouat, south of Laghouat, Hassi R'mel and Aïn-Oussera) (Gourine, Yousfi, Bombarda, Nadjemi, Stocker, et al. 2010), it was found that their activity were in a range of IC_{50} = 8.80-27.48 mg/mL. According to the mean values of IC_{50} for each location, it seems that for Laghouat and Hassi R'mel, they have approximatively the same values: IC_{50} = 17.22-18.95 mg/mL. Similar trend was observed for the sample of south of Laghouat and Aïn-Oussera with IC_{50} = 13.91-14.80 mg/mL. After the comparison with synthetic antioxidants, it was found that the essential oil of leaves *Pistacia atlantica* presents low activity in terms of scavenging free DPPH radicals.

On the other hand and for the FRAP assay applied for the same regions listed above (and refering to the same mentioned above report), it was found that the essential oil presents higher antioxidant capacity relative to the antioxidant of reference ascorbic acid. Indeed, the oil samples were nearly 3-11 times more active than ascorbic acid. The results of FRAP assay indicates that the location of Laghouat is characterized by the highest mean value of AEAC (Ascorbic acid Equivalent Antioxidant Capacity) = 9.02mg/mL. The rest of locations (south of Laghout Hassi R'mel and Aïn-Oussera) have practically the same mean values of AEAC ranging between 5.70 and 6.03 mg/mL. The antioxidant activity of essential oils of the location of Laghouat which where marked by an AEAC range values of 8.15-10.23 mg/mL and by a mean value of 9.02mg/mL, were very close to the activity of caffeic acid alue (9.65mg/mL).

In the case of the seasonal investigation of the antioxidant activity related to the location of Kheneg (Laghouat) (Gourine, Yousfi, Bombarda, Nadjemi, and Gaydou 2010), the experiment lead on of the DPPH assay shows that the value of IC_{50} during the season varies between 20.61 and 27.87 mg/mL for male essential oils and between 8.47 and 17.45 mg/mL for the female essential oils. The highest antioxidant capacity was reached in the month of June for male oils and during the months of September-October for the female oils. In general case, the female oils are more active than the male ones and especially in September and October were almost three times more active. The male oils were exhibiting a slow decreasing of their antioxidant activity all over the season (July, August, September and October Months). The same pattern was observed for the female oils from June to August; but in September and October their capacities jumped to almost twice more than their original values. The antioxidant capacity of the tested oils were weak in comparison with a chosen set of synthetic antioxidants (BHA, BHT and Ascorbic acid), which were used as positive controls.

Alternatively for the FRAP assay, the value of AEAC varied between 4.95 and 8.39 mg/mL for male oils and between 9.37 and 11.81 mg/mL for female oils. The highest activity for male oil was observed during the month of June and for the female oil it was in August. The female oils were more active than the male ones. The highest value expressing antioxidant power of the oils was compared to some synthetic antioxidants: BHA, BHT, Caffeic acid and Gallic acid. This antioxidant activity was higher than BHA and BHT; almost equal to caffeic acid but almost half the activity of Gallic acid.

Although the result of the DPPH assay shows that the essential oil has a low capacity of bleaching free radicals, the values of FRAP assay were very interesting. This important antioxidant activity reveals the promising potent of the leaves essential oil of *Pistacia atlantica* as a source of natural antioxidant.

Pituranthos Scoparius

Pituranthos scoparius (Coss. and Dur.) Schinz (*Umbelliferae*) (=*Deverra scoparia* Coss. and Dur), commonly known as "guezzah" is a Saharan species (Boutaghane et al. 2004) used in traditional medicine for the treatment of asthma and rheumatism (Boukef 1986, Benchelah et al. 2000). The Touareg people which are little communities living in the Algerian desert of Tassili and Hoggar, also use it in food as aroma for meal and bread (Benchelah et al. 2000).

Figure 6. Pituranthos scoparius.

The essential oil chemical composition of the aerial parts of *Pituranthos scoparius* were studied at a large scale in Algeria. This reports involved different regions and locations: Djelfa, Laghouat and Ghardaïa (Gourine, Merrad, et al. 2011, Vérité et al. 2004, Vernin et al. 1999). The main components identified were α-pinene (4.4-35.8%), limonene (0.8-66.5%), bornyl acetate (tr-21.0%), myristicin (tr-31.1) and dill apiole (0.4-47.3%). For limonene, it appears that the highest percentages were found in Ghardaïa area (32.7-66.5%). For the other areas, the limonene content ranges from 0.9% up to 30.0%. For myristicin and dill apiole, the variation is not clear, in some samples the content is very high and in some other ones it is very weak or nil (possible presence of chemotypes). The samples investigated presents large variations of percentage intervals relatively to each main component.

The main compounds were not the same for the overall samples analyzed and coming from the different locations i.e., the major compound is sometimes being a minor compound and vice-versa. This variability of the content of the main compounds supposes the existence of different chemotypes within the same plant.

Rhanterium Adpressum

The genus *Rhanterium* of the Inuleae tribe in the *Asteraceae* family is distributed over western North Africa (in Algeria and bordering areas of eastern Morocco), the Arabian Peninsula, Iraq and Iran (Wiklund 1986). Seven species of this genus are recognized: *R. adpressum*, *R. apressum*, *R. epapposum*, *R. eppaposum*, *R. incrassatum*, *R. squarrosum* and *R. suaveolens*. Another accepted species, *R. intermedium* Coss. and Durieu ex Pomel, is considered to be a hybrid between *R. adpressum* and *R. suaveolens* (Kala et al. 2009).

R. adpressum, commonly known in Algeria as "Arfadj", is easily recognized by its broad, densely appressed involucrol bracts. Apart from being grazed by animals, this species has also been used by the local population in the production of cheese and in folk medicine as an antidiuretic. There are only a few studies found in the literature concerning the secondary metabolites of *Rhanterium* species (Kala et al. 2009, Yaghmai and Kolbadipour 1987, Oueslati et al. 2007, Oueslati et al. 2005, Bouheroum et al. 2007, Hamia et al. 2013).

Only few investigations on the chemical composition of the essential oil of the aerial parts of *Rhanterium adpressum* Coss. and Durieu from Algeria were reported (Bouheroum et al. 2007, Hamia et al. 2013).

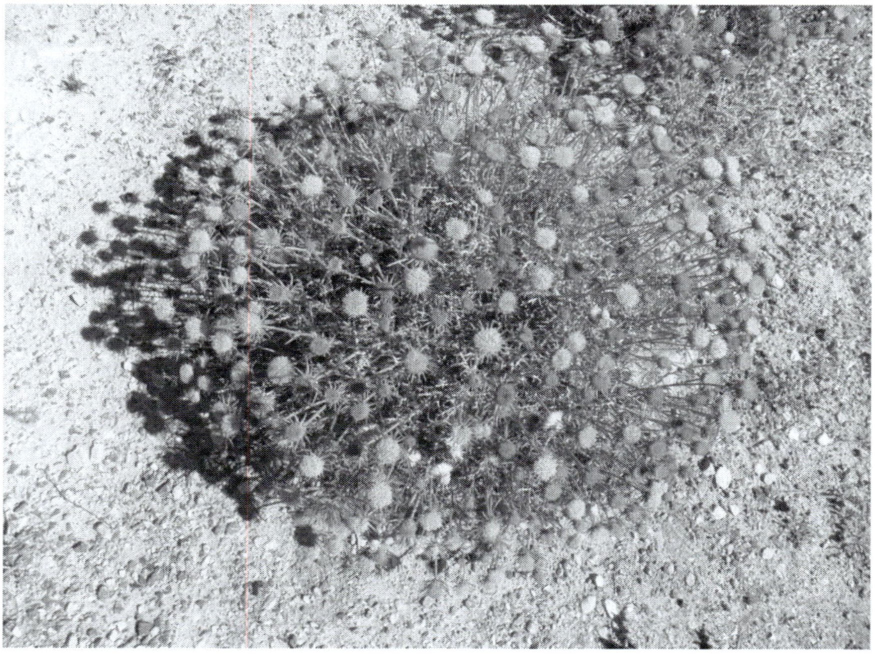

Figure 7. Rhanterium adpressum (region of Mitlili in Ghardaïa-Algeria); Photo taken in April 2012.

Figure 8. Flowers of Rhanterium adpressum (region of Mitlili in Ghardaïa-Algeria); Photo taken in April 2012.

For the essential oil of flowers; the major components identified were the monoterpene hydrocarbons: camphene (21.8%), myrcene (19.3%) and *α*-pinene (17.4%), with lower percentages of limonene (5.8%), *β*-pinene (4.5%), terpinol-4-ol (4.4%), linalool (2.5%) and sabinene (2.3%) (Hamia et al. 2013).

In contrast, the main components of the aerial parts of this plant were: spathulenol (19.6%), *β*-eudesmol (15.2%), bicyclo[4.4.0]dec-1-ene,2-isopropyl-5-methyl-9-methylene (12.9%), and *β*-cadinol (11.3%) (Bouheroum et al. 2007). These main components are completely different from those found in the flowers.

For the antioxidant activities, the value of IC_{50} from the DPPH method for the essential oil of the flowers is 24.1 mg/mL, which represent a very weak scavenging activity in comparison with the reference compound ascorbic acid (IC_{50} = 0.018 mg/mL). For the FRAP assay which measures the total antioxidant activity, it was found that the essential oil of the flowers is more active in term of reducing metal ion Fe^{++}. The results of the determined AEAC values indicate that the essential oil of the flowers (0.19 mg/mL) is presenting a fair antioxidant power which is only six times less active than vitamin E (3.1 mg/mL). Alternatively, when Phosphomolebdeum assay was performed, the essential oils were compared to another antioxiddant of reference which was ascorbic acid. In this case, the ascorbic acid was approximately 25 times more active than the essential oil.

A Proposition of a Novel Standard Approach for Expressing IC_{50} in Determining Antioxidant Activity Using DPPH Free Radical Scavenging Assay

The DPPH assay have gained in recent years a lot of interest in research paper because it offers a simple and a rapid way for evaluating the antioxidant activities of secondary metabolites of plants and vegetables. Due to the lack of a standard DPPH assay, it is difficult to compare the results reported from different research groups, and the food and nutraceutical industry cannot perform strict quality control for antioxidant products (Huang, Ou, and Prior 2005).

In her review article, Sanchez-Moreno suggested that the 2,2-di(4-*tert*octylphenyl)-1-picrylhydrazyl (DPPH) assay was an easy and accurate method with regard to measuring the antioxidant capacity of fruit and vegetable juices or extracts (Sancez-Moreno 2002).

Assessment of the Antioxidant Activity Using DPPH Assay Method

Since its first appearance in 1958, the assessment of antioxidant activity using $DPPH^{\bullet}$ (*1,1-di-phenyl-2-picrilpydrazyl*) free stable radical scavenging assay has been increasingly used as a simple way to test and evaluate the capacities of different compounds to stop or prevent the oxidation of lipids, fatty acids and other constituents of foodstuffs. This test was applied with success for synthetic food additives and natural compounds as polyphenols extracts and even recently for essential oils extracts.

A perusal of the publications in the recent past shows that various research groups have used widely different protocols which differed in the concentration of DPPH: 25 μM (Baydar,

Özkan, and Yaşar 2007), 60 μM (Prakash, Singh, and Upadhyay 2007), 100 μM (Sharififar et al. 2007), 200 μM (Xu, Chen, and Hu 2005), 300 μM (Umamaheswari et al. 2007), 500 μM (Elzaawely, Xuan, and Tawata 2007), 629 μM (Ricci et al. 2005), and many other concentrations.

As a result of these differences in reaction conditions, the IC_{50} values for even the standard antioxidants like ascorbic acid and butylated hydroxytoluene (BHT) vary a lot. Thus, it is not possible to compare the results of different laboratories. In the other hand and even though the wide spreading of the DPPH test, the determination value of the IC_{50} (or EC_{50}) using the common formula $I\% = (1-A_0/A).100\%$ is not fully accurate since it does ignore the absorbance of the total bleached free radicals (with pale yellow colour) still remaining in the tested solution. Thus the used formula overcomes the retrieving of the absorbance value (although it is weak but not negligible) of the bleached free DPPH-H radicals. This means that the classical IC_{50} value estimation is slightly underestimated. In the following paragraphs we shall give a quick and an easy way to standardise the different values of inhibition concentration obtained from different literature sources. This small adjustment contribution to the known IC_{50} value is executed by introducing a new correction factor "*f*". After the introduction of this factor *f*, which could also depends of the final concentration of the total bleached free radicals, the IC_{50} will restore its accurate value significance.

Experimentally Description of DPPH (2,2-Diphenyl-1-picrylhydrazyl) Radical Scavenging Capacity Assay

DPPH is one of a few stable and commercially available organic nitrogen radicals and has a UV-vis absorption maximum at 515 nm. Upon reduction, the solution color fades; the reaction progress is conveniently monitored by a spectrophotometer. The DPPH assay is typically run by the following procedure (Huang, Ou, and Prior 2005):

DPPH solution (3.9 mL, 25 mg/L) in methanol is mixed with sample solution (0.1 mL). The reaction progress absorbance of the mixture is monitored at 515 nm for 30 min or until the absorbance is stable. Upon reduction, the colour of the solution fades. The percentage of the DPPH remaining is calculated as

$$\%DPPH_{rem} = 100 \times [DPPH]_{rem}/[DPPH]_{t0}$$

$\%DPPH_{rem}$ is proportional to the antioxidant concentrations, and the concentration that causes a decrease in the initial DPPH concentration by 50% is defined as IC_{50}. The time needed to reach the steady state with IC_{50} concentration is calculated from the kinetic curve and defined as TEC_{50}.

Sanchez-Moreno and co-workers classified the kinetic behaviour of the antioxidant compound as follows: <5 min (rapid), 5-30 min (intermediate), and >30 min (slow). They further proposed a parameter, called "antiradical efficiency (AE)" (Jiménez-Escrig et al. 2000), to express the antioxidant capacity of a certain antioxidant. AE is calculated as:

$$AE = (1/EC_{50}) \times TEC_{50}$$

Some other recently attempts by introducing parameters like antioxidant activity index AAI (Scherer and Godoy 2009) and antioxidant activity unit (AAU) (Deng, Cheng, and Yang 2011), did not receive wide approval echo.

New Determination of the Inhibition Percentage: A More Accurate Formula

The classic formula used for the calculation (estimation) of the inhibition percentage of the bleached free radicals in DPPH assay is:

$$I\% = \left(\frac{A_0 - A}{A_0}\right)\times 100 = \left(1 - \frac{A}{A_0}\right)\times 100 \quad (1)$$

This formula is not completely accurate, since it neglects the absorption of the bleached radicals: DPPH-H.

In other terms, the calculating formula should include this adsorption value (although it is weak but not insignificant). In this condition we must introduce a correction factor "*f*" which should take in account the final absorption A_∞ of the total free radicals bleached. The new proposed formula will be expressed as follows:

$$I\% = \left[\frac{A_0 - A}{A_0}\right]\times f = \left(\frac{A_0 - A}{A_0}\right)\Big/\left(\frac{A_0 - A_\infty}{A_0}\right) \quad (2)$$

With

$$f = \frac{1}{\left(\frac{A_0 - A_\infty}{A_0}\right)} \quad (3)$$

As it should be expected, the corrected inhibition percentage value calculated using the above formula (2) is slightly higher than the one determined by the classic formula (1). The advantage of the proposed formula is that it allows to be used to standardise values of IC_{50} of previous works of literature without carrying new assays, as follows:

$$IC_{50}^{Accurate} = IC_{50} \times f$$

Finally, we could express formula (2) in another form as follows:

$$I\% = \left[\frac{A_0 - A}{A_0 - A_\infty}\right] = \frac{1-(A/A_0)}{1-(A_\infty/A_0)} \quad (4)$$

With:

$$(A_0 - A) \le (A_0 - A_\infty) \Rightarrow \frac{A_0 - A}{A_0 - A_\infty} \le 100\% \quad (5)$$

where

A_0 : is the initial absorbance

A_∞: is the final absorbance corresponding to the total bleach of the DPPH-H free radicals present in the solution

A: is the absorbance for the sample concentration

The next logical step is the experimental determination of the absorbance A_∞ as function of the initial concentration A_0.

Due to the difficulty to compare the antioxidant capacity of natural product with literature we have tried to propose a standard formula that overlaps this problem. The major problem of differences of IC_{50} values for the DPPH assay lies primarily in the fact that different initial concentrations were used for preparation of DPPH solution. So, the new proposed formula express IC_{50} in reference to an arbitrary initial DPPH concentration (equal to 100μM) and the formula is expressed as follows:

$$IC_{50}^{Standard} = IC_{50} \times \frac{A_{100}}{A_0} \quad (6)$$

where: A_{100}: is the absorbance corresponding to arbitrary initial DPPH concentration of 100μM.

In order to correctly apply this above formula, the condition of linearity between initial absorbance A_0 and initial concentration of prepared DPPH solution must be fulfilled. This condition was experimentally verified as follows. Referring to Figure 9, experimental data showed that the initial absorbance A_0 are founded to be directly proportional to the initial concentration of prepared DPPH solution (in ethanol solvent; up to A = 2.5).

Finally, some examples provided by literature were used to apply (consolidate) the proposed formula (6). In this case, different IC_{50} values for the DPPH assay for BHT antioxidant were considered and they were: 5.4 μM for an initial DPPH concentration of 25 μM (Mimica-Dukic et al. 2004); 19.8 μM for an initial DPPH concentration of 90 μM (Sokmen et al. 2004); 86.6 μM for an initial DPPH concentration of 393 μM (Ricci et al. 2005). At first sight, these IC_{50} values seem to be quiet different, because they were determined using different initial DPPH concentrations.

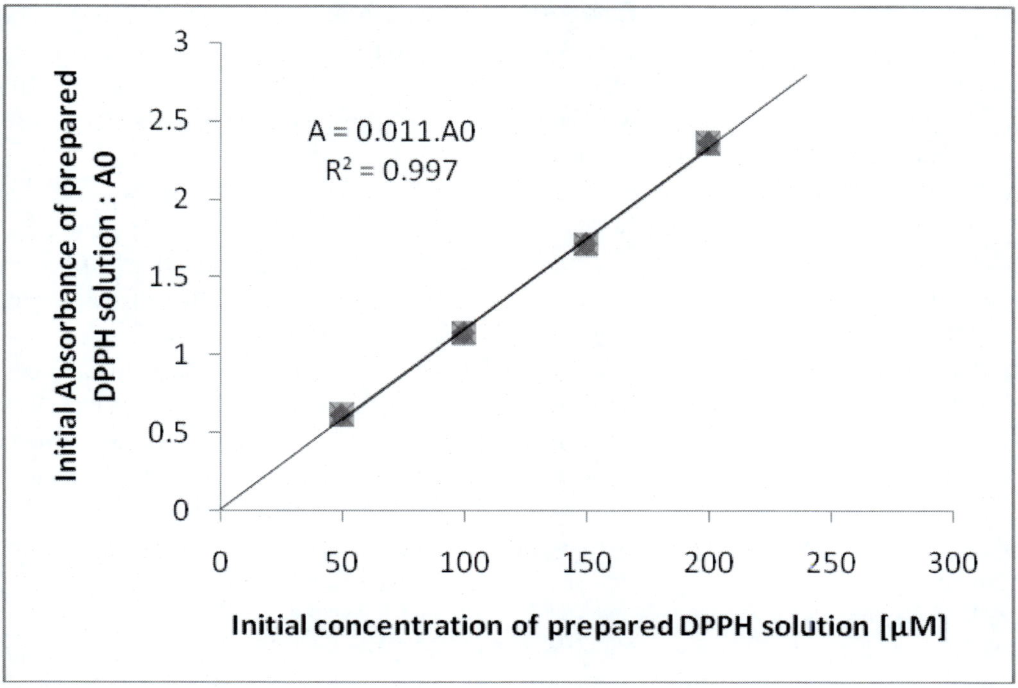

Figure 9. Initial absorbance A_0 (fresh preparation) as function of the initial concentration of prepared DPPH solution.

After correction of these values using formula (6), it was found that new IC_{50} values were respectively equal to 21.60 µM; 22.00 µM and 22.04 µM. In this case the final standardized values are very close to each other. Furthermore these values are very close to 23.81 µM founded in the works of the first and highly referenced article about DPPH assay (Brand-Williams, Cuvelier, and Berset 1995).

REFERENCES

Abdelmalek, Benabid. 2000. *Flore et écosystèmes du Maroc: Evaluation et préservation de la biodiversité.* Vol. 1. Paris: Ibis Press.

Bailey, LH. 1977. *Manual of Cultivated Plants; Most Commenly Grown in the Continental United States and Canada.* New York: Macmillan.

Barrero, AF, MM Herrador, JF Arteaga, M Akssira, F Mellouki, A Belgarrabe, and MA Blázquez. 2005. "Chemical composition of the essential oils of Pistacia atlantica Desf." *Journal of Essential Oil Research* no. 17 (1):52-54.

Baydar, Nilgün Göktürk, Gülcan Özkan, and Samim Yaşar. 2007. "Evaluation of the antiradical and antioxidant potential of grape extracts." *Food Control* no. 18 (9):1131-1136.

Bellakhdar, J. 1997. La pharmacopée marocaine traditionnelle: médecine arabe ancienne et savoirs populaires. Saint Etienne: Ibis Press.

Benchelah, Anne-Catherine, Hildegard Bouziane, Marie Maka, and Colette Ouahès. 2000. Fleurs du Sahara: voyage ethnobotanique avec les Touaregs du Tassili: Paris: Ibis Press.

Bouheroum, M, S Benayache, F Benayache, L Zaiter, JM Barrera, and L Francisco. 2007. "Terpenoids and triynepoxide from the aerial part of Rhantherium adpressum." *Chemistry of Natural compounds* no. 43 (1):110-111.

Boukef, M K. 1986. *Medecine traditionnelle et pharmacopée: Les plantes dans la médecine traditionnelle Tunisienne*. Paris, France: Agence de coopération culturelle et technique.

Boutaghane, N, A Nacer, Z Kabouche, and B Ait-Kaki. 2004. "Comparative antibacterial activities of the essential oils of stems and seeds of Pituranthos scoparius from Algerian septentrional Sahara." *Chemistry of Natural Compounds* no. 40 (6):606-607.

Brand-Williams, W, M E Cuvelier, and C L W T Berset. 1995. "Use of a free radical method to evaluate antioxidant activity." *LWT-Food Science and Technology* no. 28 (1):25-30.

Cheriti, A, N Belboukhari, K Sekkoum, and S Hacini. 2006. "Plants of Algerian semi-arid regions used for the treatment of gastro-intestinal disorders." *Journal Algérien des Régions Arides* no. 5:7-10.

Deng, Jing, Wangyuan Cheng, and Guangzhong Yang. 2011. "A novel antioxidant activity index (AAU) for natural products using the DPPH assay." *Food Chemistry* no. 125 (4):1430-1435.

Elzaawely, Abdelnaser A, Tran D Xuan, and Shinkichi Tawata. 2007. "Essential oils, kava pyrones and phenolic compounds from leaves and rhizomes of Alpinia zerumbet (Pers.) BL Burtt. and RM Sm. and their antioxidant activity." Food Chemistry no. 103 (2):486-494.

Gourine, Nadhir, Isabelle Bombarda, Mohamed Yousfi, and Emile M. Gaydou. 2010. "Chemotypes of Pistacia atlantica Leaf Essential Oils from Algeria." *Natural Product Communications* no. 5 (1):115-120.

Gourine, Nadhir, Bahia Merrad, Mohamed Yousfi, Pierre Stocker, and Emile M. Gaydou. 2011. "Chemical Composition of the Essential Oil of Pituranthos scoparius." *Natural Product Communications* no. 6 (8):1151-1154.

Gourine, Nadhir, Ibrahim Sifi, Emile M Gaydou, and Mohamed Yousfi. 2011. "Chemical Composition of the Essential Oil of Unripe Galls of Pistacia atlantica Desf. from Algeria." *The Natural Products Journal* no. 1 (2):125-127.

Gourine, Nadhir, Mohamed Yousfi, Isabelle Bombarda, Boubakeur Nadjemi, and Emile Gaydou. 2010. "Seasonal Variation of Chemical Composition and Antioxidant Activity of Essential Oil from Pistacia atlantica Desf. Leaves." *Journal of the American Oil Chemists Society* no. 87 (2):157-166.

Gourine, Nadhir, Mohamed Yousfi, Isabelle Bombarda, Boubakeur Nadjemi, Pierre Stocker, and Emile M. Gaydou. 2010. "Antioxidant activities and chemical composition of essential oil of Pistacia atlantica from Algeria." *Industrial Crops and Products* no. 31 (2):203-208.

Hamia, Chahrazed, Nadhir Gourine, Hadjer Boussoussa, Mokhtar Saidi, Emile M Gaydou, and Mohamed Yousfi. 2013. "Chemical composition and antioxidant activity of the essential oil and fatty acids of the flowers of Rhanterium adpressum." *Natural Product Communications* (8):1171-1174.

Huang, Dejian, Boxin Ou, and Ronald L Prior. 2005. "The chemistry behind antioxidant capacity assays." *Journal of Agricultural and Food Chemistry* no. 53 (6):1841-1856.

Jiménez-Escrig, Antonio, Isabel Jiménez-Jiménez, Concepción Sánchez-Moreno, and Fulgencio Saura-Calixto. 2000. "Evaluation of free radical scavenging of dietary carotenoids by the stable radical 2, 2-diphenyl-1-picrylhydrazyl." *Journal of the Science of Food and Agriculture* no. 80 (11):1686-1690.

Kala, A, N Gherraf, D Belkacemi, S Ladjel, A Zellagui, H Samir, S Chihi, and L Brahim. 2009. "Composition of the essential oil of Rhanterium adpressum Coss. and Durieu. from Algeria." *Archives of Applied Science Research* no. 1 (2):115-118.

Mecherara-Idjeri, Samira, Aicha Hassani, Vincent Castola, and Joseph Casanova. 2008. "Composition of leaf, fruit and gall essential oils of Algerian Pistacia atlantica Desf." *Journal of Essential Oil Research* no. 20 (3):215-219.

Mimica-Dukic, Neda, Biljana Bozin, Marina Sokovic, and Natasa Simin. 2004. "Antimicrobial and antioxidant activities of Melissa officinalis L. (Lamiaceae) essential oil." *Journal of Agricultural and Food Chemistry* no. 52 (9):2485-2489.

Monjauze, A. 1980. Connaissance du bétoum Pistacia atlantica Desf. *Revue Forestière Française,* 356-363.

Oueslati, M Habib, H Ben Jannet, Zine Mighri, Susan Matthew, and Pedro M Abreu. 2007. "A new C_9 nor-isoprenoid glucoside from Rantherium suaveolens." *Natural Product Research* no. 21 (10):884-888.

Oueslati, M Habib, Zine Mighri, H Ben Jannet, and Pedro M Abreu. 2005. "New ceramides from Rantherium suaveolens." *Lipids* no. 40 (10):1075-1079.

Prakash, Dhan, Brahma N Singh, and Garima Upadhyay. 2007. "Antioxidant and free radical scavenging activities of phenols from onion (Allium cepa)." *Food Chemistry* no. 102 (4):1389-1393.

Ricci, Donata, Daniele Fraternale, Laura Giamperi, Anahi Bucchini, Francesco Epifano, Giovanni Burini, and Massimo Curini. 2005. "Chemical composition, antimicrobial and antioxidant activity of the essential oil of Teucrium marum (Lamiaceae)." *Journal of Ethnopharmacology* no. 98 (1):195-200.

Sancez-Moreno, C. 2002. "Methods used to evaluate the free radical scavenging activity in foods and biologic systems-Review." *Food Science and Technology International* no. 8 (3):121-137.

Scherer, Rodrigo, and Helena Teixeira Godoy. 2009. "Antioxidant activity index (AAI) by the 2,2-diphenyl-1-picrylhydrazyl method." *Food Chemistry* no. 112 (3):654-658. doi: 10.1016/j.foodchem.2008.06.026.

Seigne, A. 1985. "La forêt circumméditerranéenne et ses problèmes." In Techniques Agricoles et Productions Méditerranéennes, 137-141. Paris: G.P. Maisonneuve et Larose.

Sharififar, F, MH Moshafi, SH Mansouri, M Khodashenas, and M Khoshnoodi. 2007. "In vitro evaluation of antibacterial and antioxidant activities of the essential oil and methanol extract of endemic Zataria multiflora Boiss." Food Control no. 18 (7):800-805.

Sifi, Ibrahim, Nadhir Gourine, Emile M. Gaydou, and Mohamed Yousfi. 2015. "Chemotypes of essential oil of unripe galls of Pistacia atlantica Desf. from Algeria." *Natural Product Research* no. 29 (20):1945-1949.

Sokmen, Atalay, Medine Gulluce, H Askin Akpulat, Dimitra Daferera, Bektas Tepe, Moschos Polissiou, Münevver Sokmen, and Fikrettin Sahin. 2004. "The in vitro antimicrobial and antioxidant activities of the essential oils and methanol extracts of endemic Thymus spathulifolius." *Food Control* no. 15 (8):627-634.

Tzakou, Olga, Ioannis Bazos, and Artemios Yannitsaros. 2007. "Volatile metabolites of Pistacia atlantica Desf. from Greece." *Flavour and Fragrance Journal* no. 22 (5):358-362. doi: 10.1002/ffj.1805.

Umamaheswari, M, K Asokkumar, R Rathidevi, AT Sivashanmugam, V Subhadradevi, and TK Ravi. 2007. "Antiulcer and in vitro antioxidant activities of Jasminum grandiflorum L." *Journal of Ethnopharmacology* no. 110 (3):464-470.

Vérité, P, A Nacer, Z Kabouche, and E Seguin. 2004. "Composition of seeds and stems essential oils of Pituranthos scoparius (Coss. and Dur.) Schinz." *Flavour and Fragrance Journal* no. 19 (6):562-564.

Vernin, Gaston, Claude Lageot, Claude Ghiglione, Mostefa Dahia, and Cyril Parkanyi. 1999. "GC/MS Analysis of the Volatile Constituents of the Essential Oils of Pituranthos scoparius (Coss et Dur.) Benth. et Hook, from Algeria." *Journal of Essential Oil Research* no. 11 (6):673-676.

Wiklund, Annette. 1986. "The genus Rhanterium (Asteraceae: Inuleae)." Botanical journal of the Linnean Society no. 93 (2):231-246.

Xu, Juan, Shubing Chen, and Qiuhui Hu. 2005. "Antioxidant activity of brown pigment and extracts from black sesame seed (Sesamum indicum L.)." *Food Chemistry* no. 91 (1):79-83.

Yaghmai, Mohammad Shahram, and Shoaleh Kolbadipour. 1987. "Volatile components of Rhanterium epapposum Oliv." *Flavour and Fragrance Journal* no. 2 (1):29-32.

In: Essential Oils
Editor: Miranda Peters

ISBN: 978-1-63484-351-5

Chapter 8

COMPARATIVE STUDY OF THE ESSENTIAL OIL EFFECTS ON THE *ASPERGILLUS FLAVUS* GROWTH

***Verônica M. S. Santos*[1], *Fábio V. Sussa*[2], *Edlayne Gonçalez*[1], *Paulo S. C. Silva*[2] *and Joana D'arc Felicio*[1,*]**

[1]Centro de Pesquisa e Desenvolvimento de Sanidade Animal, Instituto Biologico, São Paulo, Brazil

[2]Instituto de Pesquisas Energéticas e Nucleares, Universidade de São Paulo, São Paulo, Brazil

ABSTRACT

Essential oils of plant species have shown activity against a variety of pathogenic and toxigenic fungi. The toxigenic fungi can cause a lot of damage to food commodities, including mycotoxin production. The presence of molds and mycotoxins in food commodities is a potential risk to health of humans and livestock. The aflatoxins, mutagenic and carcinogenic substances, are mycotoxins produced especially by *Aspergillus flavus* and *A. parasiticus*. This work reports the comparative effects of the three essential oils leaves on growth of A. flavus and their chemical constitutions. The essential oils of *Peperomia pellucida*, *Leunurus sibiricus* and *Porophyllum ruderale* were obtained by hydrodistillation in a Clevenger-type apparatus, and their effects were evaluated in *A. flavus* culture. *A. flavus* growth was evaluated by disk diffusion assay. Filter paper disks containing 0.0, 2.5, 5.0, 7.5, and 10.0 μL of essential oils were used, and commercial fungicide was the positive control. All volumes of the three essential oils reduced the fungal growth when compared with that of the control ($p < 0.05$). The essential oil of *P. ruderale* showed the greatest inhibition effect. *L. sibiricus* essential oil showed fungicidal effect greater than that of *P. pellucida* essential oil ($p < 0.05$). The chemical composition of the essential oils used was determined by GC-MS analysis. The

[*] Corresponding author: Joana D'arc Felicio. Centro de Pesquisa e Desenvolvimento de Sanidade Animal, Instituto Biologico, Av. Conselheiro Rodrigues Alves, 1252, CEP 04014-002, São Paulo, Brazil. Telephone: +55-11-50871753, fax: +55-11-50871754, e-mail address: felicio@biologico.sp.gov.br.

use of essential oils with lower toxicity than those from synthetic products can be a good alternative to *A. flavus* control.

1. Introduction

The indiscriminate use of synthetic fungicides has led to some microorganisms developing resistance to them. Additionally, there is a need to develop new fungicides that do not impact the nature. During recent years, the use of essential oils has been a promise. Their effects have been reported in various articles exhibiting retardation of mycotoxin production as well as of fungal growth, besides having shown to be less toxic [1, 2].

Essential oils are a rich source of biologically active compounds and potential sources of novel antimicrobial and antifungal agents. Several essential oils of different family species have been related as growth inhibitor of phytopathogenic and toxigenic fungi [3].

The fungi can cause a lot of damage to the grains during planting, harvesting and storage [4]. The economic impact of fungal invasion includes the decline of the power of germination, moldering with visible discoloration, odor, chemical and nutritional changes, with consequent loss of quality, mycotoxin production, making them unfit for consumption [5]. The presence of molds and mycotoxins in food commodities is a potential health threat to humans and livestock.

Mycotoxins are toxic substances that can enter the human and animal food chain, not only by the direct ingestion of contaminated seeds or processed food, but also by the consumption of meat or other animal products (e.g., milk) from livestock fed with contaminated feed and silages [6]. Overall, there are a number of approaches that can be taken to minimize mycotoxin contamination in animal feeds and to simultaneously act in prevention of fungal growth [7]. The exploitation of natural substances with bioactivity against fungi has been the target of interest in the search for ecologically safe products [8].

Some species of *Aspergillus* are highly aflatoxigenic, i.e., they are aflatoxin producers, particularly in tropical and subtropical countries. Aflatoxins are mycotoxins produced especially by *Aspergillus flavus*, *A. parasiticus* and *A. nomius.* The aflatoxins are mutagenic and carcinogenic in animals and humans. Furthermore, aflatoxin B_1 (AFB_1) is classified as group 1 human carcinogen by the International Agency for Research on Cancer [9, 10].

A. flavus fungus can grow on a wide range of agricultural commodities, such as maize grains, peanuts and is responsible for the production of aflatoxin B_1 and B_2 (AFB_2), which are toxic metabolites contaminants of the several food commodities [11]. *A. flavus* can cause a high degree of deterioration when present on foods and can be responsible for considerable economic losses [12, 13]. Therefore, the control of *A. flavus* is extremely important for agriculture and public health.

Numerous essential oils of medicinal plants have been tested to ascertain their ability in controlling aflatoxin contamination in culture medium conditions and *Aspergillus* growth [3, 11 14, 15]. Several species of *Thymus* have been reported to possess antifungal properties against *A. parasiticus, A. terreus, A. niger, A. fumigatus* and *A. flavus*, and the aflatoxin production was also inhibited [16, 17]. *Ageratum conyzoides* essential oil inhibited the *A. flavus* growth on average 55%; on the other hand, the aflatoxin production was completely inhibited [3].

Mishra (1994) [18], during screening of essential oils isolated from different parts of 14 species of higher-order plants at 3,000 ppm, reported that only the essential oil of *Cymbopogon citratus* (family Poaceae) possessed absolute toxicity against *A. flavus*. *Cinnamomum zeilanicum* essential oil also showed inhibition of the mycelial development of *A. flavus* [19].

Pimpinella anisum L., *Peumus boldus* Mol, *Hedeoma multiflora* Benth, *Syzygium aromaticum* L., and *Lippia turbinate* essential oils were evaluated against *Aspergillus* section Flavi in sterile maize grain under different water activity (Aw) conditions. All concentrations assayed of the plants completely inhibited the fungal growth rate of the four isolates studied when Aw conditions were 0.95 and 0.90, whereas *P. anisum* and *S. aromaticum* showed only a minor inhibitory effect [20] on the growth of this fungus.

P. anisum, P. boldus, Mentha piperita, Origanum vulgare and *Minthosthachys verticillata* essential oils showed antifungal effect against *Aspergillus* section Flavi (two isolates of *A. parasiticus* and two isolates of *A. flavus*). These essential oils inhibited the percentage of germination, germ-tube elongation rate and fungal growth. AFB_1 accumulation was completely inhibited by *P. anisum, P. boldus, O. vulgare essential oils. M. piperita* and *M. verticillata* essential oils inhibited AFB_1 production by 85-90% at all concentrations assayed [21].

T. vulgaris, *Satureja hortensis* and *Syzygium aromaticum* essential oils were evaluated in culture medium and tomato paste. *A. flavus* was inoculated in Sabouraud Dextrose Broth and tomato paste. Results showed that all essential oils could inhibit the growth of *A. flavus*. Taste panel evaluations were carried out in a tomato ketchup base, and the percent of inhibition of each essential oil in tomato paste was lower than in culture medium [22].

Several species of different genus of Piperaceae, Asteraceae and Lamiaceae have shown antifungal activity against various fungi.

The genus *Peperomia*, belonging to the family Piperaceae, was described by Ruiz and Pavon [23] and comprises about 1,600 species distributed in Central America, South America, Asia, Africa and Oceania [24, 25].

P. pellucida (L.) Kunth species is a herbaceous plant, whose popular names in Brazil are 'erva-de-jabuti,' 'alfavaca-de-cobra' and 'coraçãozinho' [26]. It is traditionally known for its medicinal properties like analgesic, bactericidal, fungicidal, emollient, diuretic, anti-inflammatory and also is used against cough, headache and to treat boils, bleeding and to control cardiac arrhythmia [27-30].

P. pellucida essential oil is constituted of monoterpenes, sesquiterpenes, arylpropanoids, flavones, flavonols and phytosterols [28, 29, 31]. Fungicidal activity of *P. pellucida* essential oil is due to presence of the bioactive substances such as phenylpropanoids [32, 33].

Porophyllum ruderale (Asteraceae) is a ruderal aromatic herb, known popularly in Brazil as couve-cravinho, arnica-brasileira, arnica-do-campo. In popular medicine it is used as anti-inflammatory, antifungal, antibacterial, sedative, to combat hypertension, in the treatment of leishmaniosis, edema and injury, in the treatment of snake bites, arthritis, and pain. The healing activity has been related to the presence of varying amounts of phenolic compounds of the type tannins [34, 35].

The aerial parts of *P. ruderale* have been previously analyzed by different research groups because of their volatile compounds. The main components found in essential oil of *P. ruderale* were limonene, (*E, E*)-dodecatrienol and β-myrcene. These substances have insecticidal and fungicidal activity. Bohlmann, Jakupovic, Robinson, and King (1980) [36] reported thiophene derivatives with unsaturated chains and a chemical compound derivative from thymol in petroleum ether extracts. Besides, it is known that this essential oil can be used as a pesticide [35, 37].

A mixture of limonene and β-phellandrene (50.3%), sabinene (20.2%), 1- undecene (4.7%), 4-terpineol (3.8%), and α-pinene (2.9%) were identified in essential oil from the fresh aerial parts of *P. ruderale* collected in Venezuela. Antibacterial activity of the essential oil of this species was evaluated against gram-positive (*Staphylococcus aureus*, *Enterococcus faecalis*) and gram-negative (*Escherichia coli*, *Klebsiella pneumoniae*, *Pseudomonas aeruginosa*) bacteria using the disc diffusion agar method [38].

Leonurus sibiricus (in Brazil, erva-macaé or rubim) belongs to the family Lamiaceae, produces terpenoids and phenolic substances with allelopathic effects [39]. According to Lorenzi [40], this plant is considered invasive in agricultural fields, in almost all Brazilian territory. Its use as a medicinal herb is indicated based on folk tradition, in cases of colds, bronchitis and rheumatism [41]. Moreover, the folk medicine recommends its green parts, such as teas, in the cases of postpartum bleeding, excessive menstruation, and against edema, abscess and kidney problems; the leaves and flowers of this plant (infusion) also are capable of fighting vomiting and diarrhea [42]. *L sibiricus* hydroalcoholic extract inhibited the growth of *Candida albicans, Staphylococcus aureus* and *Pseudomonas aeruginosa* in 70% [43].

This work reports the antifungal activity, the comparative study of the effects of essential oils from *Peperomia pellucida*, *Porophyllum ruderale*, and *Leonurus sibiricus* on *A. flavus* growth and the principal chemical components of the essential oils of these species.

2. Materials and Methods

2.1. Plant Materials

Leaves of *L. sibiricus* and *P. ruderalle* were collected in the garden of the Instituto Biológico, São Paulo city, São Paulo State, Brazil, in March/2010. Samples of *P. pellucida* were collected in the Rio de Janeiro Botanical Garden, RJ, Brazil, in February 2011 and identified by Elsie F. Guimarães (Botanical Garden Research Institute, RJ, Brazil).

2.2. Oil Extraction and Analysis

The fresh leaves *L. sibiricus* and *P. ruderale* and aerial parts of *P. pellucida* were cut into small pieces and placed in a distillation Clevenger apparatus for 2 hours. The hydrolyte was extracted with hexane and evaporated at room temperature, and the resulting oil was stored in dark glass bottles in a freezer until it was used by test in *A. flavus* and GC/MS analysis.

GC-MS analyses of the main components of essential oil and its fraction were done on a Shimadzu QP-5000 equipped with an OV-5 (30 m x 0.25 mm x 0.25 μm, Ohio Valley Specialty Chemical, Inc.) capillary column. Operating conditions were undertaken at oven temperature from 60°C to 240°C at 3°C/min, injector and detector temperatures of 240°C and 230°C, respectively; system operated by electron impact ionization (70 eV), helium as a carrier gas at a constant flow of 1.7 mL/min, split 1/20. The oil components were identified using retention indices with those of authentic compounds or with literature data [44, 45].

2.3. Determination of Kováts Retention Index

A standard mixture of *n*-alkanes (C_7H_{16}-$C_{40}H_{82}$ Sigma-Aldrich, 99%) was used to verify GC/MS system performance and to calculate the Kováts index (KI) of each compound in the sample. The default (1 μL) of the alkane mixture was injected into the GC/MS system operating under the same conditions described above, and their retention times (minutes) were used as an external reference standard to calculate the K_i along with the retention times of each compound of interest [45, 46].

2.4. Antifungal Assay

2.4.1. Culture Conditions

A. flavus strain producer of aflatoxin B1 was isolated from peanut and provided by the Instituto de Biociências/USP, São Paulo, Brazil. The fungi were plated onto potato dextrose agar (PDA) and incubated for 10 days at 25°C. The spore suspension used as inoculum was prepared washing cultures with sterile 0.01% Tween 80 solution.

2.4.2. Disk Diffusion Assay

Filter paper disks (6 mm diameter) containing 2.5, 5.0, 7.5, and 10.0 μL of the three crude essential oils were applied on the dextrose agar in petri dishes previously inoculated with the fungal inoculum on the surface. The inoculated plates were incubated at 25°C for 5 days. At the end of this period, antifungal activity was evaluated by measuring the zone of inhibition (mm) against the test fungus [47]. The commercial fungicide was used as the positive control. All treatments consisted of three replicates and were repeated three times, and the averages of the experimental results were determined. Antifungal experiments were performed in triplicate, and the data were analyzed by means of one-way ANOVA.

3. RESULTS AND DISCUSSION

The hydrodistillation of leaves of *L. sibiricus* and *L. ruderale* and aerial parts of *P. pellucida* yielded pale yellow colored oil (yield: 0.03, 0.08, and 0.03%, v/v respectively). The yield of essential oil can vary considerably depending upon the source plant, location, time and period of collection. The chemical composition of the essential oils used, determined by GC-MS analysis, emphasized the presence of different major compounds (Table 1). Figures 1, 2 and 3 show the constituents of the *L. sibiricus*, *L. ruderale* and *P. pellucid*, respectively.

The chemical composition of the *L. sibiricus* essential oil consists of 14 substances, of which 12 were identified and represented for long-chain hydrocarbons 72.38%, ketones and aldehydes 12.94%, and monoterpene hydrocarbons 9.32%.

Table 1. Chemical analysis of *P. pellucida, P. ruderale*, and *L. sibiricus* essential oils

KI*	Constituent		%	
		P. pellucida	*P. ruderale*	*L. sibiricus*
902	2-heptanone			13.21
905	isocitronelene		5.85	
912	2- methyl-4-heptanone		1.24	
917	*n*-nonane			9.31
933	tetrahydrocitronelene		0.78	
937	3-methyl- 4-heptanone			2.56
941	β-citronelene		1.66	
959	6- methyl-5-hepten-2-one			5.00
961	3-octanone		11.70	
968	myrcene		1.03	
971	*n*-octanal			32.55
1094	linalool			5.69
1199/1195	*n*-dodecane	0.48		0.96
1294	*n*-tridecane			1.37
1394/1393	*n*-tetradecane	1.72	18.94	
1493/1492	*n*-pentadecane	0.70	34.43	
1413	*trans*-caryophyllene	2.62		7.68
1447	α-humulene			5.45
1456	aromadendrene	0.83		
1474	γ-muurolene	0.99		
1490	bicyclogermacrene	1.01		
1501	*(E, E)*-α-farnesene	0.57		
1555	germacrene B	20.86		
1557	*trans-n*erolidol	1.60		
1576/1573	spathulenol	0.77		1.01
1590/1576	caryophyllene oxide	10.83		7.45
1592	*n*-hexadecane		11.86	
1618	dillapiole	53.35		
1673	apiole	2.02		
1691	*n*-heptadecane		3.64	
1791	*n*-octadecane		2.56	
1890	*n*-nonadecane		0.95	
	Long-chain hydrocarbons	2.9	72.38	24.85
	Ketones and aldehydes		12.94	40.11
	Monoterpene hydrocarbons		9.32	
	Oxigenated monoterpenes			5.69
	Sesquiterpene hydrocarbons	26.90		13.13
	Oxigenated sesquiterpenes	13.20		8.46
	Aryl and phenylpropanoids	55.37		
	Total	**98.4**	**94.64**	**95.24**

*Experimental.

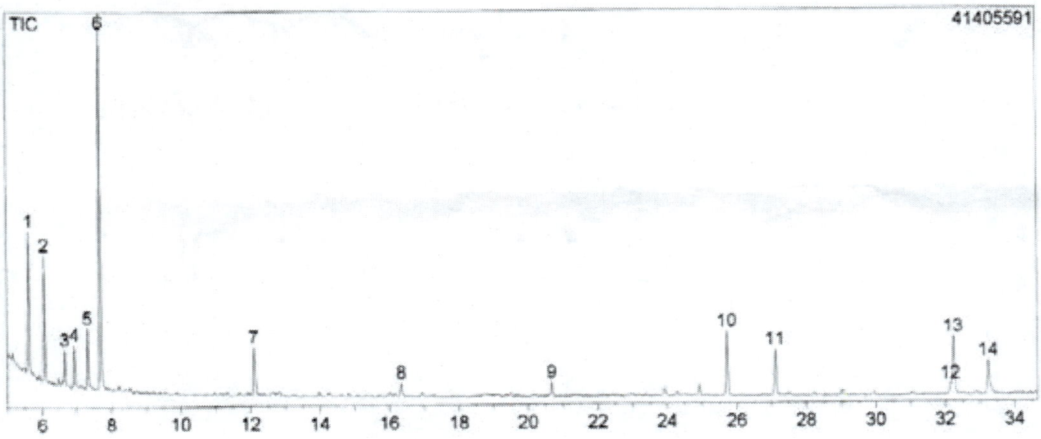

Figure 1. Chromatogram of *L. sibiricus* essential oil: (1) 2-heptanone (2) *n*-nonane, (3) 3-methyl-4-heptanone (5) methyl-5-hepten-2-one, (6) *n*-octanal (7) linalool, (8) dodecane, (9) *n*-tridecane, (10) *trans*-caryophyllene (11) α-humulene, (12) spathulenol (13) caryophyllene oxide.

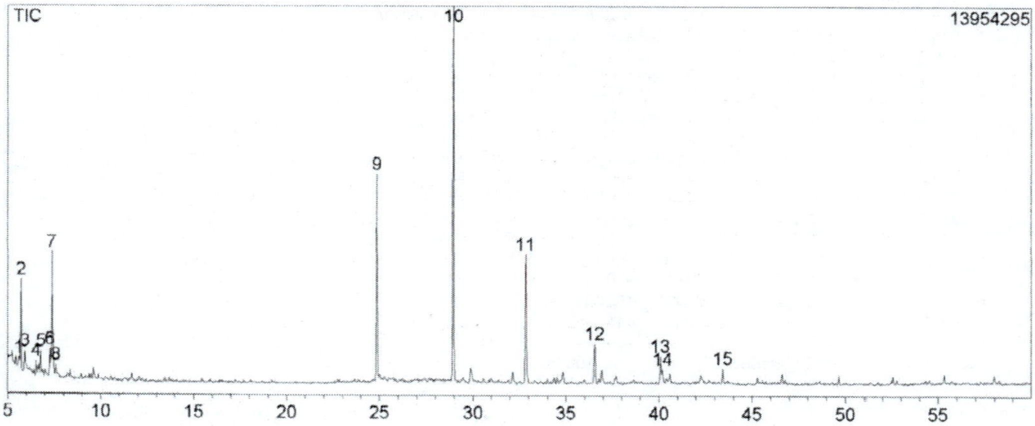

Figure 2. Chromatogram of *P. ruderale* essential oil: (2) isocitronelene, (3) 2-methyl-4-heptanone, (4) citronelene, (5) β-citronelene, (6), (7) 3-octanone, (8) myrcene, (9) *n*-tetradecane, (10) *n*-pentadecane, (11) *n*-hexadecane, (12) *n*-heptadecane, (13) *n*-octadecane, (15) *n*-nonadecane.

The main chemical constituents of the essential oil from leaves of *L. sibiricus* are *n*-octanal (32.55%), 2-heptanone (13.21%), *n*-nonane (9.31%) and *trans*-caryophyllene (7.68%), and caryophyllene oxide (7.45%). Almeida et al. (2005) [48] reported as major constituents of *L. sibiricus* essential oil from leaves the *trans*-caryophyllene (33.43%), germacrene D (24.95%), and α-humulene (21.49%). The main constituents of *P. ruderale* essential oil are hydrocarbons: *n*-pentadecane (34.43%), *n*-tetradecane (18.94%), *n*-hexadecane (11.68%) and also 3-octanone (11.70%). The essential oil composition of *P. ruderale* collected in São Paulo shows very different chemical composition of previously reported studies about this essential oil. In *P. ruderale* essential oil of leaves collected in Bolivia, sabinene was the monoterpene as the main constituent (64%) [49]; limonene is the main constituent (71.4%) of essential oil leaves collected in Mexico.

Neto et al. (1994) [50] also reported limonene as the main constituent of the essential oil of leaves originated from Ceará State, Brazil.

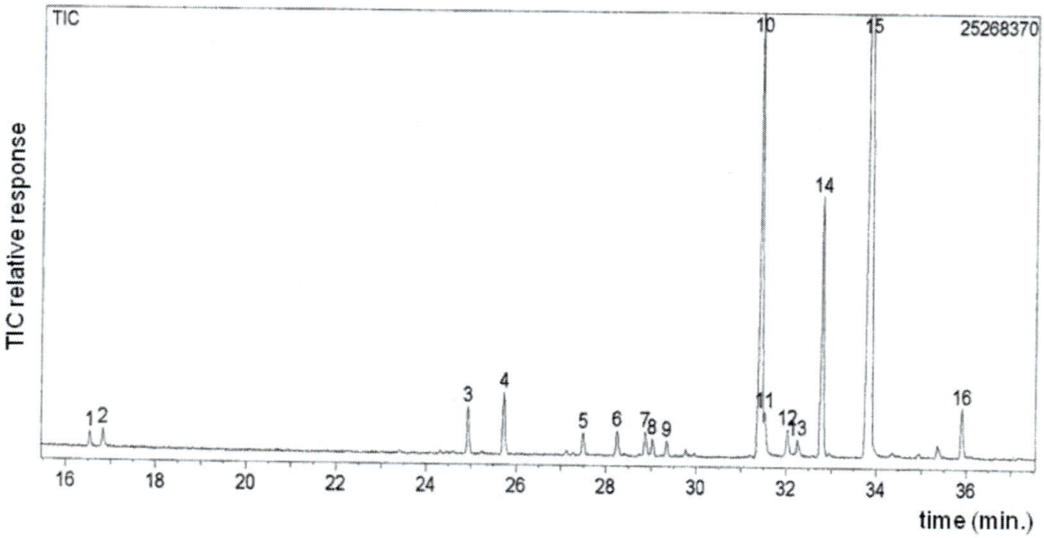

Figure 3. Chromatogram of *P. pellucida* essential oil: (1) *n*-dodecane, (2) not identified, (3) *n*-tetradecane, (4) *trans*-caryophyllene, (5) aromadendrene, (6) γ-muurolene, (7) bicyclogermacrene, (8) *n*-pentadecane, (9) (*E,E*)-α-farnesene, (10) germacrene B, (11) *trans*-nerolidol, (12) not identified, (13) spathulenol, (14) caryophyllene oxide, (15) dillapiole, (16) apiole.

The chemical composition of essential oil of *P. ruderale* has varied according to the geographical origin of the plant. In addition, the geographical differences, the different abiotic factors also influence in the chemical composition, which explains the high concentration of hydrocarbons as a constituent of essential oil studied, because the plant collected in the city of São Paulo was subjected to the stress of pollution.

The volatile compositions from the essential oil of *P. pellucida* showed arylpropanoids and sesquiterpenes as the major fractions. These results are in agreement with published data for some other *Peperomia* species [31, 51, 52, 53].

Arylpropanoids represent 55.4% and the sesquiterpenes 40.1% (non-oxygenated, 26.9%, and oxygenated, 13.2%). Long- chain hydrocarbons were found in minor amounts (2.9%), and in this work as the presence of monoterpenes was not detected.

The major constituents found in the *P. pellucida* essential oil were dillapiole (53.35%), germacrene B (20.9%), and caryophyllene oxide (10.9%). Dillapiole, which is present in a high proportion in this study, has been reported previously in the literature as the main volatile compound [52, 31]. Dillapiole has been described for its insecticidal, molluscicidal and fungicidal properties [54, 30].

Fungal growth inhibition by the disk diffusion test is very used in the evaluation of plant extracts and essential oils [32, 33]. The influence of the essential oil of the three plants on the *A. flavus* growth was measured for the volumes of 2.5, 5.0, 7.5; 10.0 μL; the inhibitory zone is in Table 2. The commercial fungicide (control) was measured at 2.19 cm.

All volumes of the essential oils of the three plants tested showed growth inhibitory effect of *A. flavus*. Inhibition of fungal growth was dependent on the volume of essential oil used, and, when compared to control, all volumes used were statically significant ($p < 0.05$). The percentage inhibition of *A. flavus* growth for all *P. ruderale* essential oil volumes were above 100%, and 10.0 μL of the oil was sufficient to completely inhibit the fungal growth (Figure

4). The volumes of 5.0 and 10.0 µL of *L. sibiricus* essential oil also inhibited the growth of *A. flavus* above 100% (Figure 4).

P. pellucida essential oil showed lower inhibitory effect on the growth of *A. flavus*. To inhibit 50% of fungal growth (IC_{50}) 5.03, 2.18 and 1.08 µL of essential oils of *P. pellucida*, *L. sibiricus*, and *P. ruderale* were necessary, respectively (Figure 5).

Several authors have reported the inhibition of fungal growth of *A. flavus* and aflatoxin biosynthesis by essential oils [3, 11, 55]. Recent articles have reported the antifungal effect of essential oils *Thymus vulgaris* L. and *Cinnamomum cassia* L. against *A. flavus* spores [56, 57, 58]. Morphological evaluation was performed by both light microscopy and scanning electron microscopy, showing that antifungal activity of essential oil of *T. vulgaris* could be detected starting at a concentration of 50 µg/mL [57].

The results in this paper showed that the fungicidal activity of the essential oils studied is quite interesting since there is no previous report of the fungicidal activity of the three essential oils against *A. flavus*. Thus, the results obtained demonstrate that essential oils are a great promise for the control of fungi, especially *A. flavus*, showing a path to sustainable agriculture.

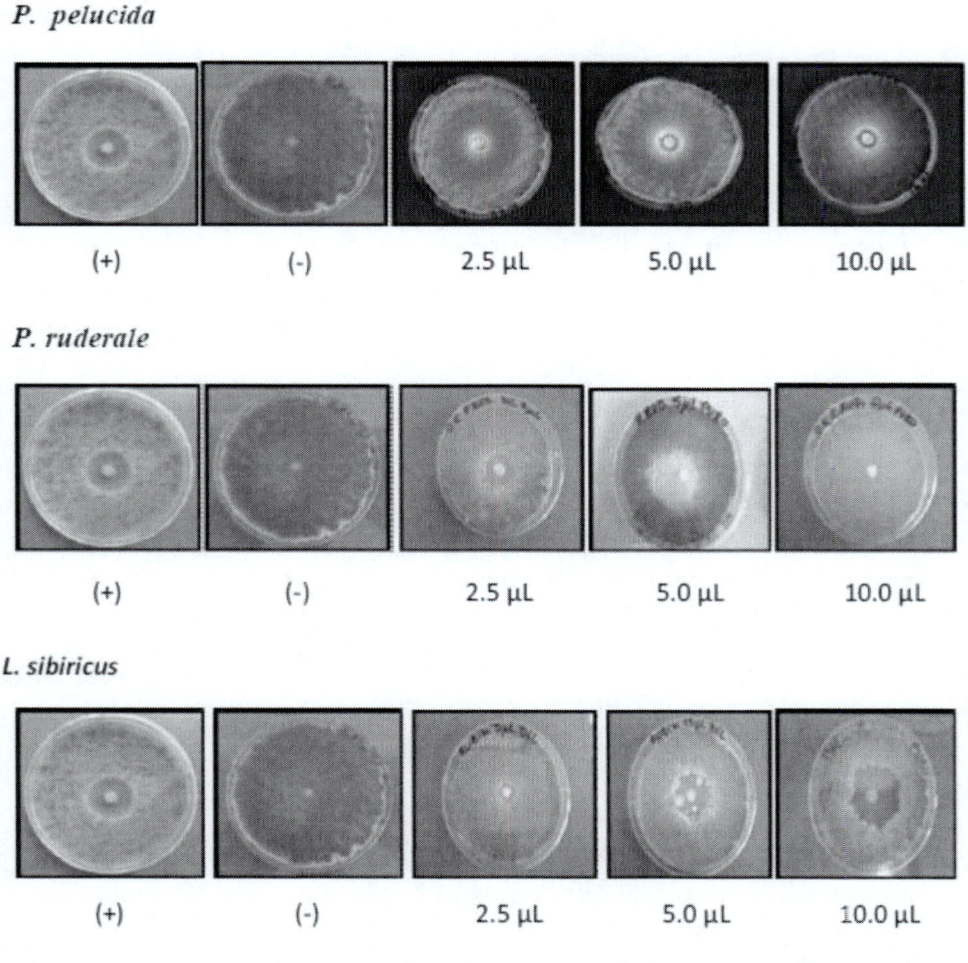

Figure 4. Inhibition zones of *A. flavus* growth by positive control (+), negative control (-), 2.5 µL, 5.0 µL, and 10.0 µL of *P. pellucida*, *P. ruderale*, and *L. sibiricus* essential oils, respectively.

Table 2. Antifungal activity of essential oils and commercial fungicide against *A. flavus*

Volumes (μL)	Inhibition zones (mm) mean ± SD*			
	P. pellucida	*P. ruderale*	*L. sibiricus*	Commercial fungicide
2.5	0.84 ± 0.50	2.4 ± 1,67	1.22 ± 0,32	
5.0	1.05 ± 0.40	4.11 ± 0,44	1.78 ± 0,90	2.19 ± 1.0
7.5	0.96 ± 0.60	4,10 ± 0,05	2.53 ± 1,40	
10	1.17 ± 1.30	Total inhibition	2.79 ± 0.95	

*Mean ± standard deviation (SD) where n = 5 means followed by a different letter are significantly different at $p \leq 0.05$, (Tukey's test).

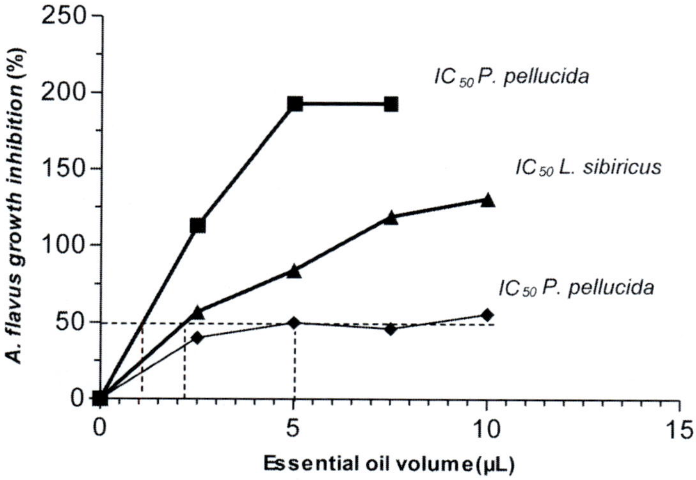

Figure 5. Curve of volumes of essential oils of *P. ruderale*, *L. sibiricus*, *and P. pellucida* versus percentage of inhibition of *A. flavus* growth.

CONCLUSION

The present study showed that the three aromatic plants from Brazil (*Peperomia pellucida*, *Leunurus sibiricus* and *Porophyllum ruderale*), used in popular medicine, exhibited variability in their essential oil content and composition. The results stress that such aromatic plants can be a new source of essential oils with an economic potential to control *A. flavus* growth, being the essential oil of *P. ruderale* the most efficient one. For this purpose, the research of the active principles of these aromatic plants can contribute to valorization of the natural resources for future cultivation, conservation, and sustainability.

REFERENCES

[1] McKenzie, K. S., Sarr, A. B., Mayura, K., Bailey, R. H., Miller, D. R., Rogers, T. D., Norred, W. P., Voss, K. A., Plattner, R. D., Kubena, L. F. and Phillips, T. D. 1997. "Oxidative degradation and detoxification of mycotoxins using a novel source of ozone." *Food Chemistry and Toxicology* 35: 807-820.

[2] Rasooli, I., Fakoor, M. H., Yadegarinia, D., Gachkar, L., Allameh, A. and Rezaei, M. B. 2008. "Antimycotoxigenic characteristics of *Rosmarinus officinalis* and

Trachyspermum copticum L. essential oils." *International Journal of Food Microbiology* 122:135-139 access in 03/15/2009 doi: 10.1016/j.ijfoodmicro.2007.11.048.

[3] Nogueira, J. H. C., Gonçalez, E., Galleti, S. R., Facanali, R., Marques, M. O. M. and Felicio, J. D. 2010. "*Ageratum conyzoides* essential oil as aflatoxin suppressor of *Aspergillus flavus*." *International Journal of Food Microbiology* 137: 55-60. access in 10/15/2010 doi: 10.1016/j.ijfoodmicro.2009.10.017.

[4] Silker, J. H. and Elliot, R. P. 1980. *"Ecologia Microbiana de los Alimentos."* [Microbial Ecology Food] Acribia, Zaragoza 1:74-96.

[5] Paster, N. and Bullerman, L. B. 1988. "Mould spoilage and mycotoxins formation in grains as controlled by physical means." *International Journal of Food Microbiology* 7: 257-265.

[6] Zjalic, S., Reverberi, M., Ricelli, A., Granito, V. M., Fanelli, C. and Fabbri, A. A. 2006. "Trametes versicolor: a possible tool for aflatoxin control." *International Journal of Food Microbiology* 107: 243-249. Access in 03/11/2009 doi: 10.1016/j.ijfoodmicro.2005.10.003.

[7] Bryden, W. L. 2012. "Mycotoxin contamination of the feed supply chain: Implications for animal productivity and feed security." *Animal feed science technology* 173: 134-158. Access in 06/22/2013 doi.org/10.1016/j.anifeedsci.2011.12.014.

[8] Passone, A. M., Girardi, N. S., Ferrand, C. A. and Etcheverry, M. 2012. "*Invitro* evaluation of five essential oils as botanical fungitoxicants for the protection of store peanuts form *Aspergillus flavus* and *A. parasiticus* contamination." *International biodeterioration and biodegradation* 70: 82-88. Access in 06/22/2013 doi: 10.1016/j.ibiod.2011.11.017.

[9] Kedia, A., Prakash, B., Mishra, P. K. and Dubey, N. K. 2014. "Antifungal and antiaflatoxigenic properties of *Cuminum cyminum* (L.) seed essential oil and its efficacy as a preservative in stored commodities." *International Journal of Food Microbiology* 168: 1-7. Access in 10/02/2015 doi: 10.1016/j.ijfoodmicro.2013.10.008.

[10] Mishra, P. K., Singh, P., Prakash, B., Kedia, A., Dubey, N. K. and Chanotiya, C. S., 2013. "Assessing essential oil components as plant-based preservatives against fungi that deteriorate herbal raw materials." *International biodeterioration and biodegradation* 80: 16-21. Access in 06/22/2013 doi:10.1016/j.ibiod.2012.12.017.

[11] Medeiros, R. T. S., Gonçalez, E., Felicio, R. C. and Felicio, J. D. 2011. "Evaluation of antifungal activity of *Pittosporum undulatum* L. essential oil against *Aspergillus flavus* and aflatoxin production." *Ciências e Agrotecnologia* 35: 71-76. Access in 1103/01/2009 dx.doi.org/10.1590/S1413-70542011000100008.

[12] Carmo, E. S., Lima, E. O. and Souza, E. L. 2008. "The potential of *Origanum vulgare* L. (Lamiaceae) essential oil in inhibiting the growth of some food related *Aspergillus* species." *Brazilian Journal of Microbiology* 39: 362-367. Access in 03/11/2009 doi.org/10.1590/S1517-83822008000200030.

[13] Ritter, A. C., Hoeltz, M. and Noll, I. B. 2011. "Toxigenic potential of *Aspergillus flavus* tested in different culture conditions." *Ciência e Tecnologia de alimentos* 31: 623-628. Access in 06/22/2013 doi.org/10.1590/S0101-20612011000300011.

[14] Omidbeygi, M., Barzegar, M., Hamidi, Z. and Naghdibadi, H. 2009. "Antifungal activity of thyme, summer savory and clove essential oils against *Aspergillus flavus* in

liquid medium and tomato paste." *Food Control* 18: 1518-1523. Access in 03/14/2011 doi:10.1016/j.foodcont.2006.12.003.

[15] Kumar, A., Shukla, R., Singh, P. and Dubey, N. K. 2010. "Chemical composition, antifungal and antiaflatoxigenic activities of *Ocimum sanctum* L. essential oil and its safety assessment as plant based antimicrobial." *Food and Chemical Toxicology* 48, 539-543. Access in 02/22/2011 doi:10.1016/j.fct.2009.11.028.

[16] Rasooli, I. and Owlia P. 2005. "Chemoprevention by thyme oils of *Aspergillus parasiticus* growth and aflatoxin production." *Phytochemistry* 66: 2851-2856.

[17] Kumar, A., Shukla, R., Singh, P., Prasad, C. S. and Dubey, N. K. 2008. "Assessment of *Thymus vulgaris* L. essential oil as a safe botanical preservative against post harvest fungal infestation of food commodities." *Innovative Food Science and Emerging Technologies* 9: 575-580. Access in 06/22/2010 doi:10.1016/j.ifset.2007.12.005.

[18] Mishra, A. K. and Dubey, N. K. 1994. "Evaluation of some essential oils for their toxicity against fungi causing deterioration of stored food commodities." *Applied and Environmental Microbiology* 60: 1101-1105.

[19] Singh, H. N. P., Prasad, M. M. and Sinha, K. K. 1993. "Efficacy of leaf extracts of some medicinal plantas against disease deve lop ment in banana." *Letters in Applied Microbiology* 17: 269-271.

[20] Bluma, R. V. and Etcheverry, M. G. 2008. "Application of essential oils in maize grain: Impact on *Aspergillus* section *Flavi* growth parameters and aflatoxin accumulation." *Food Microbiology* 25: 324-334. Access in 03/11/2009 doi: 10.1016/j.fm.2007.10.004.

[21] Bluma, R. V., Amaiden, M. R., Daghero, J. and Etcheverry, M. G. 2008. "Control of Aspergillus section Flavi growth and aflatoxin accumulation by plant essential oils." *Journal of Applied Microbiology* 105: 203-214. Access in 03/11/2009 doi: 10.1111/j. 1365-2672.2008.03741.x.

[22] Omidbeygi, M., Barzegar, M., Hamidi, Z. and Naghdibadi, H. 2007. "Antifungal activity of thyme, summer savory and clove essential oils against Aspergillus flavus in liquid medium and tomato paste." *l Food Control* 8: 1518-1523. Access in 03/11/2009 doi:10.1016/j.foodcont.2006.12.003.

[23] Ruiz, A. H. and Pavon, J. 1974. *Flora peruvianae et chilensis prodomus*. Madrid.

[24] Egwuche, R. U., Odetola, A. A. and Erukainure, O. L. 2011. "Preliminary investigation into the chemical properties of Peperomia pellucida (L.)." *Research Journal of Phytochemistry* 5: 48-53. Access in 06/22/2013 10.3923/rjphyto.2011.48.53.

[25] Xu, S., Li, N., Ning, M. M., Zhou, C. H., Yang, Q. R. and Wang, M. H. 2006. "Bioactive compounds from *Peperomia pellucida*." *Journal of Natural Products* 69: 247-250.

[26] Heinrich, M., Koehler, I., Rimpler, H. and Bauer, R. 1998 "Bioactive compounds form the mixe medicinal plant *Peperomia pellucida*." *Journal of the Mexican Chemical Society* 42: 245-248.

[27] Khan, A., Rahman, M. and Islam, S. 2007. "Antipyretic Activity of *Peperomia pellucida* Leaves in Rabbit." *Turkish Journal Biology* 32: 37-41.

[28] Bayma, J. C., Arruda, M. S. P., Müller, A. H., Arruda, A. C. and Canto, W. C. 2000. "A dimeric ArC_2 compound from *Peperomia pellucida*." *Phytochemistry* 55: 779-782.

[29] Khan, M. R. and Omoloso, A. D. 2002. "Antibacterial activity of *Hygrophila stricta* and *Peperomia pellucida*." *Fitoterapia* 73: 251-254.

[30] Arrigoni-Blank, M. F., Dmitrieva, E. G., Franzotti, E. M., Antoniolli, A. R., Andrade, M. R. and Marchioro, M. 2004. "Anti-inflamatory and analgesic activity of *Peperomia pellucida* (L.) HBK (Piperaceae)." *Journal of Ethnopharmacology* 91: 215-218. Access in 06/25/2012 doi: 10.1016/j.jep.2003.12.030.

[31] da Silva, M. H. L., Zoghbi, M. G. B., Andrade, E. H. A. and Maia, J. G. S. 1999. "The essential oils of *Peperomia pellucida* Kunth and *P. circinnata* Link var. *circinnata*." *Flavour and Fragrance Journal* 14: 312-314.

[32] Li, N., Wu, J. L., Xu, S., Zhang, G. L., Ando, M. and Wang, M. W. 2008. "Research progress on medicinal plants of *Peperomia genus*." *Recent Progress in Medicinal Plants* 20: 419-431.

[33] Ragasa, C. Y., Dumato, M. and Rideout, J. A. 1998. "Antifungal compounds from *Peperomia pellucid*." *Chemical Research Communications* 7: 54-61.

[34] Haslam, E. 1996. "Natural polyphenols (vegetable tannins) as drugs and medicines: possible modes of action." *Journal of Natural Products* 59: 205-215.

[35] Conde-Hernández, L. A. and Guerrero-Beltrán, J. A. 2014. "Total phenolics and antioxidant activity of *Piper auritum* and *Porophyllum ruderale*." *Food Chemistry* 142: 455-460. Acess 10/02/2015 doi: 10.1016/j.foodchem.2013.07.078.

[36] Bohlmann, F., Jakupovic, J., Robinson, H. and King, R. M. 1980. "A dithienylacetylene from *Porophyllum ruderale*." *Phytochemistry* 19: 2759-2760.

[37] Guillet, G., Blanger, A. and Arnason, J. T. 1998. "Volatile monoterpenes in *Porophyllum gracile* and *P. ruderale* (asteraceae): Identification, localization and insecticidal synergism with a-terthienyl." *Phytochemistry* 49: 423-429.

[38] Rondón, M. E., Delgado, J., Velasco, J., Rojas, J., Rojas, L. B., Morales A. and Carmona, J. 2008. "Chemical composition and antibacterial activity of the essential oil from aerial parts of *Porophyllum ruderale* (Jacq.) Cass. collected in Venezuela." *Ciencia* 16:5-9.

[39] Larcher, W. 2000. *Ecofisiologia vegetal*. São Carlos: Rima, p. 531.

[40] Lorenzi, H. and Matos, F. J. A. 2002. *Plantas Medicinais no Brasil: nativas e exóticas.* Instituto Plantarum. [Plants Medicinais Brazil: native and exotic. Plantarum Institute.] Nova Odessa, SP.

[41] Castellucci, S., Lima, M. I. S., Nordi, N. and Marques, J. G. W. 2000. "Plantas medicinais relatadas pela comunidade residente na estação ecológica de Jataí, município de Luís Antônio/SP: uma abordagem etnobotânica. (Medicinal plants reported by the resident community in the Ecological Station of Jataí - Luís Antônio / SP: Ethnobotany approach)" *Revista Brasileira de Plantas Medicinais* 3: 51-60.

[42] Bown, D. 1995. *The Herb Society of America: encyclopedia of herbs and their uses.* New York: Darling Kindersley Publ., p. 225.

[43] Wadt, N. S. Y., Ohara, M. T., Sakuda-Kaneko, T. M. and Bacchi, E. M. 1998. Antimicrobial activity of Leonurus *sibiricus*." *Revista brasileira de farmacognosia* 5: 167-174.

[44] McLafferty, F. W. and Stanffer, D. B. 1989. *Registry of Mass Spectral Data 1-2.* Wiley-Interscience, New York.

[45] Adams, R. P. 2001. *Identification of Essential Oil Components by Gas Cromatography/ Mass Spectroscopy*, Allured Publishing Corporation. Carol Stream, US.

[46] Viegas, M. C. and Bassoli, D. G. 2007. "Linear retention index for characterization of volatile compounds in soluble coffee using GC-MS and HP-Innowax column" *Química.*

Nova 30: 2031-2034. Access in 10/02/2015 doi.org/10.1590/ S0100-40422007000800040.

[47] Yin, M. C. and Tsao, S. M. 1999. "Inhibitory effect of seven *Allium* plants upon three *Aspergillus* species." *International Journal of Food Microbiology* 49: 49-56.

[48] Almeida, L. F. R., Delachiave, M. E. A. and Marques, M. O. M. 2005. "Composition of the essential oil of rubim (*Leonurus sibiricus*)." *Revista brasileira de plantas medicinais* 8:35-38. 06/25/2012 dx.doi.org/10.1590/S1516-05722011000200006.

[49] Loayza, I., Groot, W., Lorenzo, D., Dellacassa, E., Mondello, L. and Dugo, G. 1999. "Composition of the essential oil of *Porophyllum ruderale* (Jacq.) Cass. From Bolivia." *Flavour and Fragrance Journal* 14: 393-398. Access in 06/25/2012 doi:10.1002/(SICI) 1099-1026 (199911/12)14:6<393::AID-FFJ849>3.0.CO;2-5.

[50] Andrade-Neto, M., Bezerra, M. and Freitas, R. 2002. "The essential oil of *Porophyllum ruderale* Cass (Asteraceae)." *Journal of Essential Oil Research* 14: 14-15.

[51] dos Santos, P. R. D., Moreira, D. M., Guimarães, E. F. and Kaplan, M. A. C. 2001. "Essential oil of 10 Piperaceae species from the Brazilian Atlantic forest." *Phytochemistry* 58: 547-551.

[52] Ferreira, A. D. M., Takeara, R., Lopes, N. P. and Turatti, I. C. C. 2009. Perfil químico do óleo essencial de *Peperomia pellucida (*Chemical profile of *Piperomia pellucida* essential oil). 61ª Reunião Anual da SBPC Amazônia: Ciência e Cultura, Manaus, Amazônia.

[53] Moreira, D. L., Guimarães, E. F. and Kaplan, M. A. C. 1999. "Essential oil analysis of four *Peperomia* species (Piperaceae)." *Acta Horticulturae*. 500: 65-70.

[54] Sousa, P. J. C., Barros, C. A. L., Rocha, J. C. S., Lira, D. S., Monteiro, G. M. and Maia, J. G. S. 2008. "Toxicological evaluation of the essential oil of *Piper aduncum* L." *Revista Brasileira de Farmacognosia* 18: 217-221. Access in 06/25/2012 doi.org/10.1590/ S0102-695X2008000200013.

[55] Prakash, B., Shukla, R., Singh, P., Mishra, P. K., Dubey, N. K. and Kharwar, R. N. 2011. "Efficacy of chemically characterized *Ocimum gratissimum* L. essential oil as an antioxidant and a safe plant based antimicrobial against fungal and aflatoxin B1 contamination of spices." *Food Research International* 44: 385-390. Access in 06/25/ 2012 doi:10.1016/j.foodres.2010.10.002.

[56] Rajkovic, K., Pekmezovic M., Barac, A., Nikodinovic-Runic, J. and Arsić Arsenijević, V. 2015. "Inhibitory effect of thyme and cinnamon essential oils on *Aspergillus flavus:* Optimization and activity prediction model development." *Industrial Crops and Products* 65:7-13. Access in 10/02/2015. doi:10.1016/j.indcrop.2014.11.039.

[57] Kohiyama, C. Y., Yamamoto Ribeiro, M. M., Mossini, S. A. G., Bando, E., Bomfim, N. S., Nerilo, S. B., Rocha, G. H. O., Grespan, R., Mikcha, J. M. G. and Machinski Jr., M. 2015. Antifungal properties and inhibitory effects upon aflatoxin production of *Thymus vulgaris* L. by *Aspergillus flavus. Food Chemistry* 173: 1006-1010. Access in 10/02/ 2015. doi:10.1016/j.foodchem.2014.10.135.

[58] Pekmezovic, M., Rajkovic, K., Barac, A., Senerović, L. and Arsic Arsenijevic, V. 2015. Development of kinetic model for testing antifungal effect of Thymus vulgaris L. and Cinnamomum cassia L. essential oils on Aspergillus flavus spores and application for optimization of synergistic effect. *Biochemical Engineering Journal*, 99: 131-137. Access in 10/02/2015 doi:10.1016/j.bej.2015.03.024.

In: Essential Oils
Editor: Miranda Peters

ISBN: 978-1-63484-351-5

Chapter 9

ESSENTIAL OILS APPLICATIONS IN AGRICULTURE

Francisco Wilson Reichert Jr.[1,*]**,**
Maurício Albertoni Scariot[1]**,** ***Jéssica Mulinari***[1]**,**
Marcio Antônio Mazutti[2]**,** ***Helen Treichel***[1]
and Altemir José Mossi[1,2]
[1]Federal University of Fronteira Sul, Erechim, Brazil
[2]Federal University of Santa Maria, Santa Maria, Brazil

ABSTRACT

The overuse of chemicals in agriculture has caused several problems such as the emergence of resistant pests, environmental contamination, food residues as well as health hazards to the farmer. Insects, fungi and weeds cause losses in food production every year, by competing with crops and reducing its yield or by attacking products in post-harvest. In this way the search for new alternatives to control these agricultural pests is necessary. Essential oils are extracted from different structures of different plant species, and present compounds of the secondary metabolism of the plant that can have an effect on insects, fungi and weeds. Several bioactive compounds have already proven biological activity effect on several agricultural pests. Therefore the essential oils became a source for new products for the control of agricultural pests and help in the search for sustainable, environmentally friendly and socially fair agriculture.

INTRODUCTION

The intensive use of pesticides in agriculture has led to the emergence of several forms of resistance on insects, weeds, pathogenic fungi as well as environmental contamination and human poisoning. Given the need for new forms of control, natural products are gaining

* Email: chicowrj@gmail.com.

ground as an alternative for the control of diseases, insects and weeds (Isman, 2006; Abad et al, 2006; Singh et al., 2003; Ootani et al., 2013). Studies have proved the efficiency of plant extracts obtained from several botanical species and essential oils have excelled (Maia et al., 2015).

The oils are obtained from flowers, leaves, rhizomes and fruits, mainly through the steam distillation method. The main compounds of the composition are mono and sesquiterpenes and phenylpropanoids (Bizzo, 2009).

Phytotherapy mainly with use of essential oils has been used to combat pathogenic fungi for hundreds of years, compounds in oil are often produced in the secondary metabolism of plants and microorganisms and exert protection function (Bullerman et al., 1977; Maia et al., 2015).

Essential oils have been used in several agronomic areas, with fungistatic, insecticide and herbicide effects, killing or repelling storage insects, and hampering on their feeding or reproduction, in addition to preventing the proliferation of spoilage fungi and causing damage on weeds hampering their competitiveness (Procópio et al, 2003; Kabeh and Jalingo, 2007; Ootani et al., 2013).

Several natural compounds which have pesticide substances have been studied as the basis for the development of new products, there are compounds which already have their biological efficiency well established, among them are: Pyrethroids, nicotine, azadirachtin and rotenone (Oliveira et al., 1999; Roel, 2001). Relevant compounds are sesquiterpenes and monoterpenes with insecticidal and fungicidal activity and some level of phytotoxicity (Ootani et al., 2013).

Essential oils have shown effect on insects and fungi, in the study of Coitinho et al., (2010) the persistance of oils from *Piper hispidinervum*¸ *Melaleuca leucadendron*, *Eugenia uniflora*, *Schinus terebinthifolius* and *Piper marginatum* was evaluated on *Sitophilus zeamais* in storage corn, and it was found that the insects mortality ranged from 93.8 to 100% in the early periods of exposure, decreasing over time. Related to fungi several oils has presented effect on *Fusarium* sp.; *Aspergillus niger*, *Alternaria solani*, and others (Pereira et al., 2006; Balbi-Peña et al., 2006; Ootani et al., 2013).

Given the importance of the plant health in the cultivation of plants and the damage caused by the use of pesticides to health and the environment, essential oils are becoming a viable alternative to control pathogenic agents, agricultural pests and weeds (Ootani et al., 2013).

There is a huge botanical diversity in the world that has not been studied and this demonstrates the possible development of bioactive products from these plant resources. Since it is necessary to search for a more sustainable agriculture.

Antifungal Activity

Much of the essential oils have antimicrobial action. This activity is related to the action of certain substances in the composition of these oils such as monoterpenes, phenolics and terpenoids compounds (Gilles et al., 2010; Maia et al., 2015). The biological activity of the essential oil compounds can act with fungicidal action, depending on the tested doses (Maia et al., 2015).

Several studies are already showing the fungicidal effect of the essential oils of several plant species. In the study of Stevic et al., (2014) who tested the effect of the essential oils from *Citrus aurantium amara, Citrus aurantium bergamia, Citrus limon, Coriandrum sativum, Eucalyptus globulus, Illicium verum, Lavandula angustifolia, Matricaria recutita, Melaleuca alternifolia, Ocimum basillicum, Pelargonium graveolens, Rosa damascena, Satureja hortensis, Thymus vulgaris, Viola odorata* and *Origanum haracleoticum*, against 21 phytopathogenic fungi of pre and post-harvest found that the essential oils from *Satureja hortensis, Origanum haracleoticum, Thymus vulgaris* and *Rosa damascena* showed a high fungicidal effect compared to other tested oils, which can be the base for studies of possible products with antifungal activity. In addition, a combination of some of these oils showed higher antifungal activity compared to their single use, what can even reduce the required dose.

It is discussed that the antifungal activity of the essential oils is possibly the result of chitin infiltrating the hyphae wall, damaging the lipoprotein cytoplasmic membrane, leading to cytoplasmic leakage, as well as hyphal emptying and wilting (Zambonelli et al., 1996; Dos santos et al., 2013; Maia et al., 2015).

In the study of Rasooli et al., (2006), it was observed by electron microscopy that the *Aspergillus niger* structures exposed to 125 ppm concentration of essential oil from *Thymus eriocalyx* showed lack of cytoplasm, folding of the nuclear membrane and thickened cell wall, resulting in the growth inhibition of the fungi and leading to irreversible morphological damages.

This result corroborates with Xing et al., (2014) that, when testing the essential oils from cinnamon, eucalyptus, anise, peppermint, camphor, eugenol and synthetic chemical compounds, as cinnamaldehyde, against *Fusarium verticillioides*, showed that all tested oils in concentrations above 40 μL per petri plate (90 mm) inhibited the growth of *F. verticillioides*, highlighting the essential oil of cinnamon, that inhibited the increase in diameter and growth of *F. verticillioides*, which remained at 1.1 ± 0.1 cm from 6 to 20 days, proving to be more effective than the other tested oils. By observing in the microscope cinnamon oil effects on the fungus, it was found that *F. verticillioides* mycelia showed changes in the apical regions along the length of the hyphae. The hyphae became thin, gnarled and shrunked, in addition to the appearance of craters, damage and disruption of the cell wall.

Several plant species that have significant fungitoxic components are considered weeds or invasive plants (not native to a specific location). In the study of Eloff et al. (2007), it was concluded that the species *Amaranthus spinosus, Vigna unguiculata, Solanum mauritianum, Ricinus communis* and *Cestrum laevigatum* showed effect on micro-organisms, including *Aspergillus niger, Fusarium oxysporum, Phytophthora nicotiana* and *Rhizoctonia solani.* One reason why weeds and invasive plants are more successful is for resisting more readily to pathogen attack because these plants propagate in an independent way, and can be exposed to different ecosystems and to the attack of many pathogens, thus speeding up the process of adapting to new adversity, unlike plants of agricultural interest.

For the most part, studies showed that the essential oils of a wide range of plant species showed inhibition in the growth and proliferation of several fungal species (Table 1). This inhibition may be closely linked to several physiological disorders caused by the action of compounds present in the oils, and their effectiveness is also related to the tested dose.

Some plant species have a broad antifungal activity spectrum. Sharma and Tripathi (2006) study showed that plants of the family *Rutaceae* such as the species *Citrus sinensis* exhibited fungicidal effect on 10 pathogenic fungi of post-harvest.

Table 1. Examples of studies on inhibitory effect of essential oils on fungal species

Essential oil	Fungus	Action	Author
Thymus vulgaris *Pimpinella anisum* *Rosmarinus officinalis*	*Pyrenophora teres* *Fusarium culmorum* *Oculimacula yallundae*	Fungistatic and fungicidal	Matusinsky et al., (2015)
Cinnamomum spp.	*Fusarium verticillioides*	Fungistatic	Xing et al., (2014)
Cymbopogon citratus	*Colletotrichum gloeosporioides*	Fungistatic	Carnelossi et al., (2009)
Satureja hortensis *Satureja spicigera* *Satureja khuzistanica*	*Aspergillus niger* *Penicillium digitatum* *Botrytis cinérea*	Fungistatic and fungicidal	Forzaneh et al., (2015)
Pimenta dioica	*Aspergillus flavus* *Aspergillus fumigatus* *Penicillium brevicompactum*	Fungistatic	Zabka et al., (2009)
Origanum heracleoticum *Rosa damascena* *Eucalyptus globulus* *Satureja hortensis*	*Fusarium subglutinans* *Fusarium solani* *Fusarium tricinctum* *Penicillium* sp.	Fungistatic and fungicidal	Stevic et al., (2014)
Porophyllum linaria	*Aspergillus flavus*	Fungistatic	Juárez et al., (2015)
Tagetes patula	*Penicillium digitatum*	Fungistatic	Romagnoli et al., (2005)

Matusinsky et al., (2015) tested the essential oil of five plant species and found that all oils presented antifungal effect over five cereal infesting fungi. The effectiveness of essential oil varied according to the dose, plant species and plant pathogen. Different species of fungi showed to be more or less susceptible to essential oils from different plants, this is due to the majority compounds present in the oil. Testing 32 essential oils from different botanical species on *Aspergillus fumigatus* and *Candida krusei*, Correa-Royero et al., (2010) showed that from these 32 oils, only six had effect on *C. krusei,* four on *A. fumigatus* and seven exhibited broad fungicidal effect action on both microorganisms. This selectivity to just one micro-organism may be related to the majority compound present in the oil, being more toxic to a specific fungal species than to others. Thus, the essential oils obtained from plants are a possibility on the control of phytopathogenic fungi, because they present fungitoxic action itself or fungistatic, inhibiting the mycelial growth and spore germination, and they can induct phytoalexins (Ootani et al., 2013). This way, they are able to become a new alternative for small-scale farmers, bringing food security for the producer and less risk to the environment.

Insecticidal Activity

The occurrence of some insect species known as pests can cause losses in the field and in storage, as well as increases in the use of pesticides, as this is currently the main method of insect control in agriculture.

This way, the use of essential oils for the control of pests in agriculture has been discussed in numerous studies (Table 2). However, little is known about the action mechanisms and the active ingredient responsible for the effects they cause in insects. This fact is due to the wide range of compounds present in the essential oils, as their composition varies with plant species, and with some factors related to cultivation, soil and climate (Knaak and Fiuza, 2010).

Differences in composition and insecticidal activity of essential oils from plant species were found in the control of *Tribolium castaneum* using the essential oils from *Cymbopogon citratus* and *Eucalyptus citriodora*. The results indicated differences in composition and insecticidal activity of essential oils, demonstrating that the essential oil from *Cymbopogon citratus* caused higher mortality of insects (Olivero-Verbel et al., 2010). In the control of *Ephestia kuehniella*, *Plodia interpunctella* and *Acanthoscelides obtectus* with the use of essential oils from *Origanum onites*, *Satureja thymbra* and *Myrtus communis* was verified that the different compositions of the essential oils of the studied plant species caused distinct effects on each insect, and *Acanthoscelides obtectus* showed the lowest toxicity to the essential oils (Ayvaz et al., 2010).

Composition and insecticidal effect of essential oils can also vary between plants of the same species depending on the cultivation system. The essential oil from *Salvia aucheri* spp. *blancoana* extract from wild and cultivated plants showed differences in its composition, especially of the major compounds camphor, camphene and α-pinene. Moreover, the essential oil extracted from wild plants was more effective in the control of *Tribolium castaneum* insects (Khiyari et al., 2014).

Table 2. Examples of studies on essential oils effects in insects

Essential oil	Insect	Action	Author
Artemisia herba-alba *Artemisia absinthium*	*Orysaephilus surinamensis* *Tribolium castaneum*	Insecticide	Bachroucha et al., (2015)
Syzygium aromaticum *Cinnamomum zeylanicum*	*Sitophilus zeamais* *Acanthoscelides obtectus*	Insecticide and repellent	Correa et al., (2015) Jumbo et al., (2014)
Anethum graveolens, *Carum carvi* *Cuminum cyminum*	*Sitophilus oryzae*	Insecticide	Kim et al., (2013)
Ocimum basilicum Citrus sp.	*Sitophilus zeamais* *Tribolium castaneum*	Insecticide	Kim e Lee (2014)
Compounds: thymol and linalool	*Helicoverpa armigera* *Spodoptera litura Chilo partellus*	Insecticide and repellent	Koul et al., (2013)
Achillea biebersteinii, *A. santolina* and *A. mellifolium*	*Tribolium castaneum*	Insecticide	Nenaah (2014)
Thymus vulgaris *Pogostemon cablin* *Corymbia citriodora*	*Bemisia tabaci*	Insecticide and repellent	Yang et al., (2010)
Litsea cubeba	*Lasioderma* *Serricorne* *Liposcelis bostrychophila*	Insecticide and repellent	Yang et al., (2014)
Artemisia vestita	*Sitophilus zeamais*	Insecticide	Chu et al., (2010)

The essential oils normally cause neurotoxic effects, influencing on insects nutrition, reproduction and behaviour, resulting in death or repellency. Such effects can be related to different action mechanisms, for example by means of GABA, synapse octopamine and the inhibition of acetylcholinesterase (Regnault-Roger et al., 2012). However, the effects caused by essential oils depend on several factors, such as the way in which they come in contact with the insect, which can be by inhalation, ingestion and contact, as well as the insect species (Regnault-Roger, 1997).

The main application methods of essential oils for the insects control found on literature are fumigation, ingestion and contact. Fumigation is a method that uses the volatilization of compounds present in the essential oil, so that they come in contact with the insects by inhalation or absorption through the pores. The ingestion and contact method is usually applied as food substrate, which is treated with the essential oil and made available to the insects. Thereby, the insects do the digestion of the essential oil in addition to the topical effect, so that there is absorption by pores and by inhalation (Hashemi and Safavi, 2012; Yazdgerdian et al., 2015).

Franz et al., (2011) verified that essential oils of *Cymbopogon citratus*, *Zingiber officinale* and *Mentha* sp. applied by fumigation and contact resulted in mortality rates around 85% and greater than 90%, respectively, in the larger doses tested on *Sitophilus oryzae*. Likewise, Chu et al., (2010) found insecticidal effects of *Artemisia vestita* essential oil on *Sitophilus zeamais* with contact and fumigation applications. In another study, Saroukolai et al., (2010) found that *Thymus persicus* essential oil applied by fumigation showed insecticidal activity against *Sitophilus oryzae* and *Tribolium castaneum*.

Liu et al., (2010) found insecticide potential of essential oils of *Artemisia capillaris* and *Artemisia mongolia* on *Sitophilus zeamais* insects when applied by fumigation and contact. Also, Kiran and Prakash (2015) found that the essential oil from *Rosmarinus officinalis* applied by fumigation on *Sitophilus oryzae* and *Oryzaephilus surinamensis* insects caused toxicity and antifeedant action, and reduced by around 40% the acetylcholinesterase enzyme activity. Likewise, Yang et al., (2012) found that the essential oil from *Allium sativum*, as well as two of its major components, diallyl disulfide and the diallyl trisulfide, caused toxicity, inhibition of oviposition and behavioral change in *Sitotroga cerealella* insects.

In another study, Emamjomeh et al., (2014) found that the essential oil from *Zataria multiflora* has insecticide potential on *Ephestia kuehniella* when applied by fumigation. And the studies of Aref et al., (2015) demonstrated that the essential oil from *Eucalyptus dundasii* provided, besides toxicity by fumigation, effect on feeding and behavior of *Rhyzopertha dominica* and *Oryzaephilus surinamensis*.

The repellent action of essential oils on insects is not yet well understood. However it is known that it results from the individual and combined action of some compounds, such as α-pinene, limonene, citronellol, citronellal, camphor and thymol (Nerio et al., 2010). Caballero-Gallardo et al., (2011) found repellent action of essential oils from *Tagetes lucida*, *Lepechinia betonicifolia*, *Lippia alba*, *Cananga odorata* and *Rosmarinus officinalis*, as well as their major compounds applied alone, including limonene and α-pinene on *Tribolium castaneum*.

Licciardello et al., (2013) found repellent action of essential oils from *Cymbopogon nardus*, *Origanum vulgare* and *Rosmarinus officinalis* on *Tribolium castaneum* when applied to repellent films used as food packing. And Bougherra et al., (2015) found repellent action of *Pistacia lentiscus* essential oil on *Rhyzopertha dominica*, *Sitophilus zeamais* and *Tribolium confusum*. In other studies, Mishra et al., (2012) found repellent action of essential oils from

Eucalyptus globulus and *Ocimum basilicum* on *Tribolium castaneum* and *Sitophilus oryzae*. However, repellent action of some compounds may not be effective for long because they are extremely volatile. The studies of Zhang et al., (2011) showed that limonene and citronellal compounds isolated from the essential oil from *Cymbopogon distans* reduced their repellent ability 4 hours after the application to *Tribolium castaneum* insect. The action provided by essential oils on insects such as mortality and repellency, can also be variable in relation to applied dose and exposure time of insects to the essential oil. The studies of Abdel-Sattar et al., (2010) demonstrated that the essential oil from *Schinus molle* caused mortality and repellency effects on *Tribolium castaneum* and *Trogoderma granarium* in accordance with the increase of the applied dose and exposure time. Similarly, Zapata and Smagghe (2010) found that increasing the dose and time of exposure of essential oils from *Laurelia sempervirens* and *Drimys winteri* on *Tribolium castaneum*, higher rates of mortality and repellency were achieved.

The repellent, insecticide and behavioral action exerted by the compounds present in the essential oils extracted from plants, makes these up as a technique for insect control in agriculture. Moreover, the huge variety of species with insecticide potential as well as the enormous range of compounds, provides a wide range of options for replacement of pesticides in agriculture. However, more studies are needed in order to better elucidate the mechanisms of action and the best methods of application of essential oils for insect control.

HERBICIDAL ACTIVITY

The occurrence of weeds in cultivated areas can cause losses in productivity, reduce of product quality, can make it difficult to harvest and they can serve as an inoculums source for diseases. In addition, the need to search for alternative products, in view of the frequent cases of resistance by weeds to pesticides and environmental contamination, are extremely important for sustainable weed management in agriculture. In this way, studies are being conducted with the aim to better understand the herbicide activity of essential oils for weed control (Table 3).

The herbicidal activity of essential oils extracted from plants is characterized by allelopathy. Allelopathy occurs by the release of chemicals to the environment that can be positive or negative to the target (Inderjit, 2003). Some of the deleterious effects caused by essential oils in plants are the retardation or inhibition of germination and development of seedlings and organs (Dudai et al., 1999; Singh et al., 2003; Li et al., 2011).

Kaur et al., (2010) found that the essential oil from *Artemisia scoparia* had herbicidal effect on *Achyranthes aspera*, *Cassia occidentalis*, *Parthenium hysterophorus*, *Echinochloa crus-galli* and *Ageratum conyzoides*. The essential oil action caused a decrease in the emergence and growth of root and shoot of the plants when applied in pre-emergence, whereas when applied post-emergence, in 6 weeks old weed, caused chlorosis followed by necrosis and wilting of the plants.

The studies of Amri et al., (2012) demonstrated that the essential oil from *Pinus pinea* when in higher concentrations completely inhibited the germination of *Sinapis arvensis*, *Lolium rigidum* and *Raphanus raphanistrum* seeds, whereas when applied at lower concentrations reduced the germination, as well as the germination speed and seedling growth. In another study, Mutlu et al., (2010) found that the essential oil from *Nepeta meyeri* caused inhibition of germination and growth of ten species of weeds (*Amaranthus retroflexus, Portulaca olerace, Bromus danthoniae, Agropyron cristatum, Lactuca serriola, Bromus tectorum, Bromus intermedius, Chenopodium album, Cynodon dactylon, Convolvulus arvensis*).

The herbicide action of 27 monoterpenes on *Raphanus sativus* and *Lepidium sativum* was studied by De Martino et al., (2010), who found that these compounds cause inhibition of germination and growth of plants and may thus be used in the development of new herbicides. In another study, Chowhan et al., (2013) found inhibitory effect on germination and growth of roots and shoots of *Phalaris minor*, *Echinochloa crus-galli* and *Cassia occidentalis* by the action of the oxygenated β-pinene monoterpene, which showed that this compound acts on the integrity of cells plasma membrane. Also, Barton et al., (2010) found that 1.8 cineol, major compound of eucalyptus essential oil, caused herbicidal effect on *Lolium rigidum* and *Raphanus sativus*.

Essential oils emerge as an alternative to weed control with the use of pesticides, since several substances in their compositions have action on the germination and development of plants. In addition, several studies already address the mechanism of action of some compounds, which can in future help in the weed management. However, little is known about the selectivity of these compounds on the crops of economic interest, a fact that hampers a faster adoption of this control method.

Table 3. Recent studies on the herbicidal effects of essential oils

Essential oil	Weed	Action	Author
Pinus nigra ssp. *Laricio*	*Trifolium campestre* *Sinapis arvensis*	Inhibition of germination and growth reduction	Amri et al., (2014)
Drimys brasiliensis	*Barbarea verna* *Echinochloa crus-galli* *Ipomoea grandifolia*	Reduction of initial growth	Anese et al., (2015)
Peumus boldus	*Portulaca oleracea*	Germination inhibition	Blázquez and Carbó (2015)
Dracocephalum kotschyi	*Amaranthus retroflexus* *Chenopodium album*	Inhibition of germination and growth of the seedlings	Jalaei et al., (2015)
Artemisia scoparia	*Achyranthes asper,* *Cassia occidentalis* *Parthenium hysterophorus* *Echinochloa crus-gall* *Ageratum conyzoides.*	Reduction in emergence and seedling development, chlorosis followed by necrosis and wilting in plants with 6 weeks of age	Kaur et al., (2010)
Cymbopogon citratus	*Echinochloa crus-galli*	Inhibition of germination and early development, wilting and action on the photosynthetic process in plants with 28 days of age	Poonpaiboonpipat et al., (2013)

CONCLUSION

Currently the uses of essential oils in agriculture are restricted in their majority to research. However, prospects for the future use of these compounds are very good, as numerous studies show its deleterious effects on fungi, insects and weeds. The diversity of plants with potential for essential oils extraction, as well as the diverse range of compounds, put essential oils as great alternatives to the use of pesticides, enabling then the pursuit of a sustainable agriculture, socially and environmentally correct.

REFERENCES

Abad, M. J.; Ansuategui, M.; Bermejo, P. (2006). Active antifungal substances from natural sources. *Journal and online education inchemistry,* v. 2007, p.116-145.

Abdel-Sattar, E.; Zaitoun, A. A.; Farag, M. A.; Gayed, S. H. E., Harraz, F. M. H. (2010). Chemical composition, insecticidal and insect repellent activity of *Schinus molle* L. leaf and fruit essential oils against *Trogoderma granarium* and *Tribolium castaneum*. *Natural Product Research*, v. 24, p. 226–235.

Amri, I.; Gargouri, S.; Hamrouni, L.; Hanana, M.; Fezzani, T.; Jamoussi, B. (2012). Chemical composition, phytotoxic and antifungal activities of *Pinus pinea* essential oil. *Journal of Pest Science*, v. 85, p. 199–207.

Amri, I.; Hanana, M.; Jamoussi, B.; Hamrouni, L. (2014). Essential oils of *Pinus nigra* J.F. Arnold subsp. *laricio* Maire: Chemical composition and study of their herbicidal potential. *Arabian Journal of Chemistry* in press doi: 10.1016/j.arabjc.2014.05.026.

Anese, S.; Jatobá, L. J.; Grisi, P. U.; Gualtieri, S. C. J.; Santos, M. F. C.; Berlinck, R. G. S. (2015). Bioherbicidal activity of drimane sesquiterpenes from *Drimys brasiliensis* Miers roots. *Industrial Crops and Products*, v. 74, p. 28–35.

Aref, S. P.; Valizadegan, O.; Farashiani, M. E. (2015). *Eucalyptus dundasii* Maiden essential oil, chemical composition and insecticidal values against *Rhyzopertha dominica* (F.) and *Oryzaephilus surinamensis* (L.). *Journal Of Plant Protection Research*, v. 55, p. 35-41.

Ayvaz, A.; Sagdic, O.; Karaborklu, S.; Ozturk, I. (2010). Insecticidal activity of the essential oils from different plants against three stored-product insects. *Journal of Insect Science*, v. 10, p. 1-13.

Bachroucha, O.; Ferjanib, N.; Haouelc, S.; Jemâac, J. M. B. (2015). Major compounds and insecticidal activities of two Tunisian *Artemisia* essential oils toward two major coleopteran pests. *Industrial Crops and Products*, v. 65, p. 127–133.

Balbi-Peña, M. I.; Becker, A.; Stangarlin, J. R.; Franzener, G.; Lopes, M. C.; Schwanestrada, K. R. F. (2006). *Alternaria solani* control in tomatoes by extracts of *Cúrcuma longa* and curcumina I. *In vitro* Avaliation. *Fitopatologia Brasileira,* v. 31, p.10-14.

Barton, A. F. M.; Dell, B; Knight, A. R. (2010). Herbicidal activity of cineole derivatives. *Journal of Agricultural and Food Chemistry*, v. 58, p. 10147–10155.

Bizzo, H. R.; Hovell, A. M. C.; Rezende, C. M. (2009). Essential oils in Brazil: general aspects, development and perspectives. *Quimica Nova*, v.32, p.588-594.

Blázquez, M. A. and Carbó, E. (2015). Control of *Portulaca oleracea* by boldo and lemon essential oils indifferent soils. *Industrial Crops and Products*, v. 76, v. 515–521.

Bougherra, H. H.; Bedini, S.; Flamini, G.; Cosci, F.; Belhamel, K.; Conti, B. (2015). Pistacia lentiscus essential oil has repellent effect against three major insect pests of pasta. Industrial *Crops and Products*, v. 63, p. 249–255.

Bullerman, L. B.; Lieu, Y.; Sier, S. A. (1977). Inhibition of growth and aflatoxin production by cinnamon and clove oils, cinnamic aldehyde and eugenol. *Journal of Food Science*, v.46, p.1107- 1109.

Caballero-Gallardo, K.; Olivero-Verbel, J.; Stashenko, E. E. (2011). Repellent activity of essential oils and some of their individual constituents against *Tribolium castaneum* Herbst. *Journal of Agricultural and Food Chemistry*, v. 59, p. 1690–1696.

Carnelossi, P. R.; Schwan-Estrada, K. R. F.; Cruz, M. E. S.; Itako, A. T.; Mesquini, R. M. (2009). Essential oils in postharvest control of *Colletotrichum gloeosporioides* in papaya. *A Revista Brasileira de Plantas Medicinais*, v.11, p.399-406.

Chowhan, N.; Singh, H. P.; Batish, D. R.; Kaur, S.; Ahuja, N.; Kohli, R. K. (2013). β-Pinene inhibited germination and early growth involves membrane peroxidation. *Protoplasma*, v. 250, p. 691–700.

Chu, S. S.; Liu, Q. R.; Liu, Z. L. (2010). Insecticidal activity and chemical composition of the essential oil of *Artemisia vestita* from China against *Sitophilus zeamais*. *Biochemical Systematics and Ecology*, v. 38, p. 489–492.

Coitinho, C. B. L. R.; Oliveira, V. J.; Gondim Junior, C. G. M.; Camara, G. A. C. (2010). Essential oils persistence in storage corn, submitted to maize weevil infestation. *Revista Ciência Rural*, v.40, p.1492-1496.

Correa, Y. D. C. G.; Faroni, L. R. A.; Haddi, K.; Oliveira, E. E.; Pereira, E. J. G. (2015). Locomotory and physiological responses induced by clove and cinnamon essential oils in the maize weevil *Sitophilus zeamais*. *Pesticide Biochemistry and Physiology* in press doi:10.1016/j.pestbp.2015.06.005.

Correa-Royero, J.; Tangarife, V.; Durán, C.; Stashenko, E.; Mesa-Arango, A. (2010). *In vitro* antifungal activity and the cytotoxic effects of essential oils and extracts of medicinal and aromatic plants against *Candida krusei* and *Aspergillus fumigatus*. *Revista Brasileira de Farmacognosia,* v.20, p.734-741.

De Martino, L.; Mancini, E.; Almeida, L.F.R.; De Feo, V. (2010). The antigerminative activity of twenty-seven monoterpenes. *Molecules*, v.15, p. 6630-6637.

Dos Santos, G. R.; Brum, S. C. B. R.; Castro De, G. H.; Gonçalves, G. C.; Fidelis, R. R. (2013). Essential oils of medicinal plants effects over the helminthosporium of the Guinea grass. *Revista Ciência Agronômica*, v.44, p.587-593.

Dudai, N.; Poljakoff-Mayber, A.; Mayer, A. M.; Putievsky, E.; Lerner, H. R. (1999). Essential oils as allelochemicals and their potential use as bioherbicides. *Journal of Chemical Ecology*, v. 25, p. 1079-1089.

Eloff, J. N.; Mdee, L. K.; Masoko, P. (2007). Invasive and weedy species can be used as a source of antifungal compounds to control plant fungal pathogens. South African Association of Botanists Congress Cape Town. *South African Journal of Botany*. v.73, p.287.

Emamjomeh, L.; Imani, S.; Talebi, K.; Moharramipour, S.; Larijani, K. (2014). Chemical composition and insecticidal activity of essential oil of *Zataria multiflora* Boiss. (Lamiaceae) against *Ephestia kuehniella* (Lepidoptera: Pyralidae). *European Journal of Experimental Biology*, v. 4, p. 253-257.

Forzaneh, M.; Kiani, H.; Sharifi, R.; Reisi, M.; Hadian, J. (2015). Chemical composition and antifungal effects of three species of Satureja (S. hortensis, S. spicigera, and S. khuzistanica) essential oils onthe main pathogens of strawberry fruit. *Postharvest Biology and Technology,*v. 109, p.145-151.

Franz, A. R., Knaak, N.; Fiuza, L. M. (2011). Toxic effects of essential plant oils in adult *Sitophilus oryzae* (Linnaeus) (Coleoptera, Curculionidae). *Revista Brasileira de Entomologia*, v. 55, p. 116–120.

Gilles, M.; Zhao, J.; An, M.; Agboola, S. (2010). Chemical composition and antimicrobial properties of essential oils of three Australian *Eucalyptus* species. *Food Chemistry*. v.119, p.731-737.

Hashemi, S. M. and Safavi, S. A. (2012). Chemical constituents and toxicity of essential oils of oriental arborvitae, *Platycladus orientalis* (L.) Franco, against three stored-product beetles. *Chilean Journal of Agricultural Research*, v. 72, p. 188-194.

Inderjit, S. O. D. (2003). Ecophysiological aspects of allelopathy. *Planta*, v. 217, p. 529–539.

Isman, M. B. (2006). Botanical Insecticides, Deterrents, and Repellents in Modern Agriculture and an increasingly regulated World. *Annual Review of Entomology,* v.51, p.45-66.

Jalaei, Z.; Fattahi, M.; Aramideh, S. (2015). Allelopathic and insecticidal activities of essential oil of *Dracocephalum kotschyi* Boiss. from Iran: A new chemotype with highest limonene-10-al and limonene. *Industrial Crops and Products*, v. 73, v. 109–117.

Juárez, N. Z.; Hernández, R. L.; Bach, H.; Arreola-Sánchez, E.; Bach, H. (2015). Antifungal activity of essential oils extracted from Agastache mexicanassp. xolocotziana and Porophyllum linaria against post-harvest pathogens. *Industrial Crops and Products*, v.74, p.178-182.

Jumbo, L. O. V.; Faroni, L. R. A.; Oliveira, E. E.; Pimentel, M. A.; Silva, G. N. (2014). *Industrial Crops and Products*, v. 56, p. 27–34.

Kabeh, J. D. and Jalingo M. G. D. S. S. (2007). Pesticidal effect of bitter leaf plant *Vernonia amygdalina* (Compositae) leaves and pirimiphosmethyl on larvae of *Callosobruchus maculates* (Coleoptera: Bruchidae) and *Sitophilus zeamais* (Coleoptera: Curculionidae). *International Journal of Agriculture and Biology*, v. 9, p.452-454.

Kaur, S.; Singh, H. P.; Mittal, S.; Batish, D. R.; Kohli, R. K. (2010). Phytotoxic effects of volatile oil from *Artemisia scoparia* against weeds and its possible use as a bioherbicide. *Industrial Crops and Products*, v. 32, p. 54–61.

Khiyari, M. E. A.; Kasrati, A.; Jamali, C. A.; Zeroual, S.; Markouk, M.; Bekkouche, K.; Wohlmuth, H.; Leach, D.; Abbad, A. (2014). Chemical composition, antioxidant and insecticidal properties of essential oils from wild and cultivated *Salvia aucheri* subsp. *blancoana* (Webb. and Helder)), an endemic, threatened medicinal plant in Morocco. *Industrial Crops and Products*, v. 57, p. 106–109.

Kim, S.I.; Lee, D.W. (2014). Toxicity of basil and orange essential oils and their components against two coleopteran stored products insect pests. *Journal of Asia-Pacific Entomology*, v. 17, p. 13–17.

Kim, S.W.; Kang, J.; Park, I.K. 2013. Fumigant toxicity of Apiaceae essential oils and their constituents against *Sitophilus oryzae* and their acetylcholinesterase inhibitory activity. *Journal of Asia-Pacific Entomology*, v. 16, p. 443–448.

Kiran, S. and Prakash, B. (2015). Toxicity and biochemical efficacy of chemically characterized *Rosmarinus officinalis* essential oil against *Sitophilus oryzae* and *Oryzaephilus surinamensis*. *Industrial Crops and Products*, v. 74, p. 817–823.

Knaak, N. and Fiuza, L. M. (2010). Plants essential oils potentials in the control of insects and microorganisms. *Neotropical Biology and Conservation*, v. 5, p. 120-132.

Koul, O.; Singh, R.; Kaur, B.; Kanda, D. (2013). Comparative study on the behavioral response and acute toxicity ofsome essential oil compounds and their binary mixtures to larvae of *Helicoverpa armigera*, *Spodoptera litura* and *Chilo partellus*. *Industrial Crops and Products*, v. 49, p. 428– 436.

Li, J.; Liu, X.; Dong, F.; Xu, J.; Li, Y.; Shan, W.; Zheng, Y. (2011). Potential allelopathic effects of volatile oils from *Descurainia sophia* (L.) Webb ex Prantl on wheat. *Biochemical Systematics and Ecology*, v. 39, p. 56–63.

Licciardello, F.; Muratore, G.; Suma, P.; Russo, A.; Nerín, C. (2013). Effectiveness of a novel insect-repellent food packaging incorporating essential oils against the red flour beetle (*Tribolium castaneum*). *Innovative Food Science and Emerging Technologies*, v. 19, p. 173–180.

Liu, Z. L.; Chu, S. S.; Liu, Q. R. (2010). Chemical Composition and Insecticidal Activity against *Sitophilus zeamais* of the Essential Oils of *Artemisia capillaris* and *Artemisia mongólica*. *Molecules*, v. 15, p. 2600-2608.

Maia, F. T.; Donato De, A.; Fraga, E. M. (2015). Review: Antigungal activity of essential oils of plants. *Revista Brasileira de Produtos Agroindustriais,* v.17, p.105-116.

Matusinsky, P.; Zouhar, M.; Pavela, R.; Novy, P. (2015). Antifungal effect of five essential oils against importante pathogenic fungi of cereals. *Industrial Crops and Products*, v.67, p.208-215.

Mishra, B. B.; Tripathi, S. P.; Tripathi, C. P. M. (2012). Repellent effect of leaves essential oils from *Eucalyptus globulus* (Mirtaceae) *and Ocimum basilicum* (Lamiaceae) against two major stored grain insect pests of Coleopterons. *Nature and Science*, v. 10, p. 50-54.

Mutlu, S.; Atici, Ö.; Esím, N. (2010). Bioherbicidal effects of essential oils of *Nepeta meyeri* Benth. on weed spp. *Allelopathy Journal*, v. 26, p. 291-300.

Nenaah, G. E. (2014). Chemical composition, toxicity and growth inhibitory activities ofessential oils of three *Achillea* species and their nano-emulsions against *Tribolium castaneum* (Herbst). *Industrial Crops and Products*, v. 53, p. 252– 260.

Nerio, L. S.; Olivero-Verbel, J.; Stashenko, E. (2010). Repellent activity of essential oils: A review. *Bioresource Technology*, v. 101, p. 372–378.

Oliveira, J. V.; Vendramin, J. D. And Haddad, M. L. (1999). Bioactivity of plants powder over bean weevil in storage grains. *Revista Agrícola*, v. 74, p.217-227.

Olivero-Verbel, J.; Nerio, L. S.; Stashenko, E. E. (2010). Bioactivity against *Tribolium castaneum* Herbst (Coleoptera: Tenebrionidae) of *Cymbopogon citratus* and *Eucalyptus citriodora* essential oils grown in Colombia. *Pest Management Science*, v. 66, p. 664–668.

Ootani, A. M.; Aguiar, W. R.; Ramos, C. C. A.; Brito, R. D.; Silva Da, B. J.; Cajazeira, P. J. (2013). Use of Essential Oils in Agriculture. *Journal of biotechnology and biodiversity.* v. 4, p.162-174.

Pereira, M. C.; Vilela, G. R.; Costa, L. M. A. S. (2006). Inhibition of fungal growth using essential oils of spices. *Ciência Agrotecnológica*, v. 30, p.731-733.

Poonpaiboonpipat, T.; Pangnakorn, U.; Suvunnamek, U.; Teerarak, M.; Charoenying, P.; Laosinwattana, C. (2013). Phytotoxic effects of essential oil from *Cymbopogon citratus* and its physiological mechanisms on barnyardgrass (*Echinochloa crus-galli*). *Industrial Crops and Products*, v. 41, p. 403– 407.

Procópio, S. O.; Vendramim, J. D.; Ribeiro Júnior, J. I.; Santos, J. B. (2003). Bioactivity of various powders of vegetable origin against *Sitophilus zeamais* Mots. (Coleoptera: Curculionidae). *Ciência e Agrotecnologia*, v. 27, p.1231-1236.

Rasooli, I.; Rezaei, B. M.; Allameh, A. (2006). Growth inhibition and morphological alterations of *Aspergillus niger* by essential oils from *Thymus eriocalyx* and *Thymus x-porlock*. *Food control*, v.17, p.359-364.

Regnault-Roger, C. (1997). The potential of botanical essential oils for insect pest control. *Integrated Pest Management Reviews*, v. 2, p. 25–34.

Regnault-Roger, C.; Vincent, C.; Arnason, J. T. (2012). Essential Oils in Insect Control: Low-Risk Products in a High-Stakes World. *Annual Review of Entomology*, v. 57, p.405–424.

Roel, A. R. (2001). Use of plants with insecticidal properties: a contribution to sustainable rural development. *Revista International Desenvolvimento Local*, v. 1, p 43-50.

Romagnoli, C.; Bruni, R.; Andreotti, E.; Rai, M. K.; Vicentini, C. B.; MARES, D. (2005). Chemical characterization and antifungal activity of essential oil of capitula from wild Indian *Tagetes patula* L. *Protoplasma*. v.225, p.57-65.

Saroukolai, A. T.; Moharramipour, S.; Meshkatalsadat, M. H. (2010). Insecticidal properties of *Thymus persicus* essential oil against *Tribolium castaneum* and *Sitophilus oryzae*. *Journal of Pest Science*, v. 83, p. 3–8.

Sharma, N.; Tripathi, A. (2006). Fungitoxicity of the essential oil of *Citrus sinensis* on postharvest pathogens. *World Journal of Microbiology and Biotechnology*, v.22, p.587–593.

Singh, H. P., Batish, D. R., Kohli, R. K. (2003). Allelopathic interactions and allelochemicals: new possibilities for sustainable weed management. *Critical Reviews in Plant Sciences*. v. 22, p.239-311.

Stevic, T.; Beric, T.; Savikin, K.; Sokovic, M.; Godevac, D.; Dimkic, I.; Stankovic, S. (2014). Antifungal activity os selected essential oils against fungi isolated from medicinal plant. *Industrial Crops and Products*, v.55, p.116-122.

Xing, F.; Hua, H.; Selvaraj, N. J.; Zhao, Y.; Zhou, L.; Liu, X.; Liu, Y. (2014). Growth inhibition and morphological alterations of *Fusarium verticillioides* by cinnamon oil and cinnamaldehyde. *Food Control*, v.46, p.343-350.

Yang, F. L.; Zhu, F.; Lei. C. L. (2012). Insecticidal activities of garlic substances against adults of grain moth, *Sitotroga cerealella* (Lepidoptera: Gelechiidae). *Insect Science*, v. 19, p. 205–212.

Yang, K.; Wang, C. F.; You, C. X.; Geng, Z. F.; Sun, R. Q.; Guo, S. S.; Du, S. S.; Liu, Z. L.; Deng, Z. W. (2014). Bioactivity of essential oil of *Litsea cubeba* from China and its main compounds against two stored product insects. *Journal of Asia-Pacific Entomology*, v. 17, p. 459–466.

Yang, N. W.; Li, A. L.; Wan, F. H.; Liu, W. X.; Johnson, D. (2010). Effects of plant essential oils on immature and adult sweetpotato whitefly, *Bemisia tabaci* biotype B. *Crop Protection*, v. 29, p. 1200-1207.

Yazdgerdian, A. R.; Akhtar, Y.; Isman, M. B. (2015). Insecticidal effects of essential oils against woolly beech aphid, *Phyllaphis fagi* (Hemiptera: Aphididae) and rice weevil,

Sitophilus oryzae (Coleoptera: Curculionidae). *Journal of Entomology and Zoology Studies*, v. 3, p. 265-271.

Zabka, M.; Pavela, R.; Slezakova, L. (2009). Antifungal effect of *Pimenta dioica* essential oil against dangerous pathogenic and toxinogenic fungi. *Industrial Crops and Products*, v.30, p.250-253.

Zambonelli, A.; Aulerio, A.Z.; Bianchi, A.; Albasini, A. (1996). Effects of essential oils on phytopathogenic fungi in vitro. *Journal of Phytopathology*. v.144, p.491 -494.

Zapata, N. and Smagghea, G. (2010). Repellency and toxicity of essential oils from the leaves and bark of *Laurelia sempervirens* and *Drimys winteri* against *Tribolium castaneum*. *Industrial Crops and Products*, v. 32, p. 405–410.

Zhang, J. S.; Zhao, N. N.; Liu, Q. Z.; Liu, Z. L.; V, S. S.; Zhou, L.; Deng, Z. W. (2011). Repellent constituents of essential oil of *Cymbopogon distans* aerial parts against two stored-product insects. *Journal of Agricultural and Food Chemistry*, v. 59, p. 9910–9915.

In: Essential Oils
Editor: Miranda Peters
ISBN: 978-1-63484-351-5

Chapter 10

ESSENTIAL OILS AS GROWTH PROMOTERS IN BROILER CHICKEN

L. Azevedo[1,*], E. R. Martins[1], A. C. Almeida[1], W. C. L. Nogueira[1], D. E. Faria Filho[2] and F. S. A. Fonseca[1]

[1]Instituto de Ciências –Agrárias (ICA) – UFMG, Avenida Universitária, Bairro Universitário – Montes Claros – MG, Brasil
[3]Faculdade de Zootecnia e Engenharia de Alimentos (FZEA) – USP, Pirassununga – SP, Brasil

ABSTRACT

The large production of poultry is in part due to antimicrobials that are used to improve broiler chicken performance, although in recent years this practice has been questioned due to suspected appearance of residues in the meat and resistant microorganisms. Thus, the essential oils emerge as promising substitutes to the usual growth promoters.

The aim of this study was to analyze the state of the art concerning the use of essential oils as additives in broiler diets and such data will be used to conduct further studies in the future.

The databases used were SciELO, Portal Capes, Science Direct and PubMed. 42 papers published between 2005 and 2014 were selected. 27 plant species were tested, and oregano was the most used. The essential oils act in different ways in the organism of the animals, going beyond the antimicrobial activity and showing effect on several productive parameters of the poultry, with results similar or better than those of antimicrobials.

* Email address: azevedobella71@yahoo.com.br.

INTRODUCTION

The Brazilian poultry industry emerged around the 1940's and gradually developed in the following years until the 70s, when there was a restructuring and consequent substantial growth of the sector. The increase in the urban population, among other factors, increased demand for processed chicken meat, and the implementation of the integration system and new technologies in genetics, nutrition, management and sanitation increased supply (Santos et al. 2005; Pinotti and Paulillo, 2006).

However, in order to increase production it was necessary to overcome the susceptibility to pathogenic micro-organisms the animals presented. The development of these bacteria in the intestine causes damage both by competition for food, as well as through damage to the mucosa. To overcome these problems, antimicrobial agents have come to be added to animal feed, for prevention, at doses below those employed for the treatment of diseases (subtherapeutic). Antibiotics act by manipulating the microbiota, ensuring good animal performance and expression of the whole genetic potential. The antimicrobial agents with this function were called growth promoters and, more recently, have been referred to as performance enhancers (Santos et al. 2005).

Growth promoters prevent the development of pathogenic microorganisms in the intestine of treated animals (may inhibit or eliminate), resulting in better digestion and absorption of various essential nutrients and a more stable equilibrium in the microbial population, increasing weight gain, improving feed conversion and decreasing mortality. In general, they act promoting selectivity to beneficial microorganisms in the intestine, although the action mechanism is still not well understood (Santos et al. 2005).

However, over the years there was a concern with the likely side effects that could result from the use of antimicrobials, driven by the development of microorganism resistance in relation to some antibiotics used in human health. Antibiotics in allowed animal feed are those that are not used in the treatment of diseases in humans. However, there are cases where the microorganism that has developed resistance to an antimicrobial agent may also exhibit this characteristic against other bases (cross-resistance), resulting in drug ineffectiveness (Santos et al. 2005; Souza and Silva, 2008).

The literature contains various works denouncing the presence of antibiotic residues and resistant bacteria in products derived from chicken. 60% of the chicken meat samples analyzed by Wouafo et al. (2010) showed *Salmonella* serotypes, almost 90% of them resistant. Furtula et al. (2010) conducted experiments with diets containing growth promoters and when analyzing poultry litter samples, they found residues of antibiotics and multiresistant *Escherichia coli*. Diarra et al. (2007) also tested different promoters and found that all isolated individuals of *E. coli* had some degree of resistance to multiple antibiotics. Osman and Hariri (2013) isolated samples of *Clostridium perfringens* from poultry with necrotic enteritis and showed that all the individuals were resistant to most commonly used antibiotics.

On the other hand Thibodeau et al. (2008) evaluated the effect of Zinc bacitracin and Virginiamycin on the emergence of resistant bacteria at a commercial of broiler chicken breeder. *E. coli* and *Enterococcus* spp. were isolated and tested for the presence of resistance genes. The results showed that the use of growth promoters did not significantly influence the emergence of resistance in this case.

This concern with food security led to the ban on the use of antimicrobials as feed additives in the European Union. In Brazil, however, the use of certain bases is still permitted, though there are many discussions about it. For example, Avilamycin, Zinc bacitracin and Tylosin sulfate are allowed (although they are banned in Europe) and Chloramphenicol, systemic Sulfonamides, Tetracyclines and Oxitetraciclinas are prohibited (Souza and Silva, 2008).

Europe therefore, as a way to protect its consumers, has created new barriers to the entry of poultry products from other countries, making it difficult for Brazilian exports. This means that these producers must adapt to consumer demands to be competitive in the market. However the non-use of growth promoters has a number of difficulties as consequences. Problems arise, from the drop in performance of the animals to increased mortality, resulting in higher production costs and higher cost for consumers (Souza and Silva, 2008).

Therefore, there is a need to seek alternatives, which has boosted studies involving the use of natural additives, such as essential oils, in the role of performance enhancers (Koaiyama et al. 2014).

In this context, this paper aims to review the literature to analyze the state of the art relating to the use of essential oils as additives in the diet of broiler chickens and with such data submit new studies in future addressing the topic.

METHODOLOGY

A quantitative literature review was carried out through the collection of scientific articles published between 2005 and 2014 in four online databases: SciELO, Science Direct, Portal Capes and PubMed. Thirty-six keywords related to the subject, in Portuguese, Spanish and English were entered in the search field. They were: additives; natural additives; alternative; antibiotics; antimicrobials; birds; poultry; carcass; coacervation; performance; diets; antimicrobial effect; encapsulation; herbal extracts; plant extracts; herbal medicines; broiler chickens; intestinal histology; performance enhancer; secondary metabolites; intestinal microbiota; microencapsulation; intestinal morphology; poultry nutrition; essential oils; volatile oils; medicinal plants; growth promoter; residues in meat; bacterial resistance; cross-resistance; spray drying; replacement; subtherapeutic and supplementation.

The articles were evaluated according to the following inclusion criteria:

1) it is a scientific article;
2) covering at least one of the following subjects:
    a) use of antibiotics as a performance enhancer;
    b) resistance of microorganisms to antimicrobial agents and/or residues in meat;
    c) microencapsulation;
    d) use of essential oils from any plant against bacteria isolated from poultry *in vitro* and/or in place of antimicrobials *in vivo*.

Studies dealing with the use of essential oils as additives were organized in spreadsheets, identifying the database where they were was found; title; author; publication year; plant species used; the use (or not) of microencapsulation; type (*in vitro* or *in vivo*); results and conclusion (positive or negative).

RESULTS AND DISCUSSION

Sixty-eight works were found, of which only forty-two met the inclusion criteria. Three of the works had the use of antibiotics as performance enhancers as their central theme; five dealt with the problems with resistance of micro-organisms; only seven addressed microencapsulation, and finally twenty-seven dealt directly with using EOs as additives.

Among the articles, the plant species used varied, with many articles dealing with testing more than one oil simultaneously. Four articles did not mention the plants used, referring to them only as "blends of essential oils."

Oregano was the most cited plant (17 articles), followed by thyme and cinnamon (7), rosemary (5), pepper, sage and clove (4), garlic (3), citrus peel, bay leaves, basil, fennel and ginger (2 each?) and lemon, peppermint, fenugreek, eucalyptus, cumin, copal, cinnamon-of-china, boldo-do-chile, red pepper tree, star anise and rosemary pepper (1 each ?). All these species showed positive results as growth promoters, individually or in association. The main components of the essential oils of these plants are carvacrol and thymol, cinnamaldehyde, eugenol, capsaicin and allicin, but it is not definitively known if the other components present in the oils acted synergistically with the active ingredients.

Plant extracts may have bactericidal or bacteriostatic effect, acting similarly to antibiotics. Their action is due to their chemical components, called active ingredients. These are secondary metabolites produced by plants in order to protect it in adverse situations - stress, environmental factors, predator and pathogen attack - but such metabolites also provide them with medicinal properties. Glycosides, alkaloids, phenolic and polyphenolic compounds, terpenes, saponins, mucilage and flavonoids are some common secondary metabolites (Barreto et al. 2008).

The essential oils are secondary metabolites that are part of the terpene or terpenoid class. They are characterized by their low molecular weight and consequent volatility. Their function in the plant typically consists of attracting pollinators or repel predators because of their strong odor. Their formation is influenced by several factors (time of year, temperature, humidity, exposure to sun and wind) especially when the storage structures of these compounds, of the species concerned, are located in the leaf (Barreto et al. 2008; Bona et al. 2012). The collection time also affects the essential oil characteristics, therefore it is recommended to collect the morning or evening to avoid exposure to the sun (Melo et al. 2011), in addition to observe the time of year, which can affect the plant biomass production and oil yield (Figueiredo et al. 2009).

The action mode essential oils is not yet fully understood. What is suggested is that these substances may increase the palatability of the diet, stimulate the secretion of endogenous enzymes, facilitate digestion, alter intestinal microflora and help in reducing sub-clinical infections, as well as exert antimicrobial activity via cell wall attack and alter the permeability of the cytoplasmic membrane of bacteria and stimulate antioxidant and nitrogen uptake (Bona et al. 2012; Traesel et al. 2011).

Several studies have demonstrated the effectiveness of essential oils *in vitro* and *in vivo*. In the study of Santurio et al. (2007) the essential oils from oregano (*Origanum vulgare*), thyme (*Thymus vulgaris*) and cinnamon (*Cinnamomum zeylanicum*) showed antimicrobial activity against strains of *Salmonella enterica* of poultry origin. The results show that oregano and thyme essential oils are effective against *Salmonella enterica*.

Barreto et al. (2008) found similar performance and organ morphometry parameters for poultry fed oregano, cloves, cinnamon and red pepper extract (microencapsulated extracts with 20% essential oil) and those fed Avilamycin. Only treatment with bell pepper presented worse morphometry results than those of the control.

Scheuermann et al. (2009) show that the use of essential oils (commercial mixture, plus capsaicins and saponins) in broiler diets improves efficiency in the use of calcium, phosphorus, energy and amino acids. This makes it possible to reduce these levels in the feed and thereby reduce costs.

According to Silva et al. (2009) oregano has an effect similar to that of Avilamycin + Salinomycin regarding the intestinal morphology of animals challenged with *Eimeria tenella*.

The essential oils of oregano, and garlic (individually or associated) can replace growth promoters, showing similar effects on performance variables and *Clostridium* counts (Kirkpinar et al. 2010).

Rizzo et al. (2010) worked with three combinations of phytogenic additives: 1) plant extracts of cloves, thyme, cinnamon and pepper; 2) synthetic essential oils of oregano and cinnamon and pepper extract oleoresin; 3) eucalyptus oil, chinese cinnamon essential oil-, boldo, Chile bilberry leaves and fenugreek seeds, compared to antibiotics. The use of the plant extracts did not differ from results obtained from the use of antibiotics for performance, carcass characteristics and use of protein and dietary energy.

Santurio et al. (2011) analyzed the *in vitro* antimicrobial effect of essential oils from the spices - oregano (*Origanum vulgare*), thyme (*Thymus vulgaris*), cinnamon (*Cinnamomum zeylanicum*), Mexican oregano (*Lippia graveolens*), ginger (*Zingiber officinale)*, sage (*Salvia officinalis*), rosemary (*Rosmarinus officinalis*) and basil (*Ocimum basilicum)* - against *E. coli* samples isolated from poultry and cattle. The authors concluded that oregano, Mexican oregano, thyme and cinnamon oils were effectively bactericidal and oregano oil presented the greatest antimicrobial activity.

Silva et al. (2011) found that 0.4% of red mastic essential oil promotes improvement in intestinal absorptive surface of broilers. Evaluating the intestinal mucosa characteristics is important because some microorganisms can cause injury. In such cases there are alterations in the mucosa structure (villi and crypts) that harm cellular turnover and nutrient absorption, impairing animal performance.

Traesel et al. (2011) concluded that the essential oils of oregano (*Origanum vulgare*), sage (*Salvia officinalis*), rosemary (*Rosmarinus officinalis*) and pepper (*Capsicum frutescens*) extract used to substitute antimicrobials showed antioxidant activity, providing lower plasma lipid peroxidation and therefore less oxidative damage in broilers.

Working with essential oils of oregano, rosemary and a mixture of oils *in vitro* and *in vivo*, Mathlouthi et al. (2012) noticed that these oils have activity against the most common bacteria of poultry and it is possible to use them to substitute the usual antibiotics, since they an effect similar to that of Avilamycin.

Good results were found by Bonn et al. (2012) with the essential oils of oregano, rosemary, cinnamon and pepper extract in the control of *Clostridium perfringens* in poultry challenged with *Eimeria* or *Salmonella Enteritidis.* The oils promoted control of bacteria, reduction in lesions, increase in the villus/crypt ratio and the number of CD3+ (T-lymphocytes) cells in the duodenum of the animals.

Agostini et al. (2012) showed that 100 and 200 mg/kg of essential oil of clove improves feed efficiency and can modulate the microflora and intestinal epithelium.

Amerah et al. (2012) used xylanase and a mixture of essential oils (containing thymol and cinnamaldehyde), associated or individually, and showed improvement in performance and a 77% reduction of *Salmonella* vertical transmission. The best results occurred in the treatment in which the xylanase and the oil blend were associated.

Betancourt et al. (2012a) obtained higher body weight and economic efficiency, besides a lower negative impact on the weight of broilers challenged with *Coccidia* receiving essential oil of rosemary-pepper.

In another study, Betancourt et al. (2012b) tested several varieties of oregano, analyzing ileal digestibility of protein, fat and energy in addition to intestinal and histomorphology performance variables. Treatments with oregano were inferior to those with the antibiotic only in the ileal digestibility of protein, showing that it may be used as a promoter. They also perceived that there is a negative correlation between the consumption of carvacrol and body weight, and another correlation, positive, with the consumption of thymol.

Erhan et al. (2012) found similar performance parameters among chickens fed diets with antibiotics and diets with essential oil of pennyroyal. There was also a decrease in the *E. coli* count and increase in the number of beneficial intestinal bacteria.

The association of essential oils of oregano, fennel and citrus peel presents high potential to replace antibiotics, improving performance and health status (improved feed efficiency and survival rate, reduced cholesterol, improved HDL concentration, did not affect bacteria in the ileum, reduced ammonia in the ileum and increased villus height in the duodenum) as well as meat quality (Hong et al. 2012).

Siva et al. (2012) used oregano essential oil and ginger extract compared to antibiotics (Bacitracin and Colistin) and found similar performance parameters and intestinal morphology for all groups.

The essential oil of peppermint and prebiotics (fructooligosaccharides) were compared with virginiamycin (growth promoter) by Emami et al. (2012). Most of the variables worsened with the use of essential oils and prebiotics, which does not make the essential oils in question good alternatives to antibiotics. Wirth et al. (2012) concluded that oregano and thyme can be used as additives improving the performance of animals tested.

The essential oils of thyme and star anise were tested by Cho et al. (2013) and were seen to promote performance improvement, reduction in total blood cholesterol and inhibition of *C. perfringens* and *E. coli* proliferation in the small and large intestine, achieving better results than those of Avilamycin.

Karadas et al. (2014) showed improved performance and improvement in hepatic concentrations of carotenoids and coenzymes in chickens supplemented with a commercial blend of essential oils (including carvacrol, cinnamaldehyde, and Capsicum oleoresin).

Khattak et al. (2014), using a commercial blend of essential oils based cumin, bay leaves, basil, lemon, oregano, sage and thyme, achieved better performance parameters and carcass yield.

The essential oils of oregano and garlic improve the chicken meat characteristics and do not change the carcass yield compared to antibiotics (Kirkpinar et al. 2014).

Table 1. Species of plants tested, sensitive and resistant microorganisms, activity as performance enhancer and primary action of essential oils in the organism

Species	Antimicrobial activity[1]	No activity[2]	*in vivo* Activity	Principal action mechanisms
Rosemary	*Salmonella, E. coli, Listeria, Clostridium, Eimeria*	*	Yes	Hinders meat oxidation; performance; carcass yield[3]
Rosemary-pepper	*	*	Yes	Performance
Garlic	*Clostridium*	*	Yes	Performance; carcass yield; hinders oxidation and increases succulence of the meat
Star anis	*E. coli, Clostridium*	*	Yes	Performance; cholesterol levels
Mastic	*	*	Yes	Performance; intestinal morphometry
Chile Bilberry	*	*	Yes	Performance
Cinnamon	*Salmonella, Clostridium, Eimeria*	*E. coli*	Yes	Performance; carcass yield
Chinese cinnamon	*	*	Yes	Performance
Citrus peels	*	*	Yes	Performance; cholesterol levels
Cumin	*	*	Yes	Performance; carcass yield
Copaiba	*	*	Yes	Performance; carcass yield
Clove	*	*	Yes	Performance; carcass yield; modifications in microbiota
Fennel	*	*	Yes	Performance; cholesterol levels
Eucalyptus	*	*	Yes	Performance
Fenugreek	*	*	Yes	Performance

Koaiyama et al. (2014) tested three combinations of essential oils: 1) rosemary, cloves, genbibre and oregano; 2) cinnamon, sage, white thyme and copaiba oleoresin; and 3) association with 50% of each of the combinations on performance and carcass yield. Mixtures 2 and 3 (association) had potential as a performance enhancer, improving weight gain, feed conversion and carcass and cut yields.

Studies show that essential oils used as additives, alone or in association, promote similar results and sometimes higher results than the antimicrobial growth promoters in various variables. These data demonstrate the feasibility of replacing such additives. Among the works assessed, there was only one work in which the results were negative (Emami et al. 2012), in which the authors suggest further studies, since the use of peppermint, in the form of essential oil, had not been reported in the literature, plus the fact that the variation of environmental conditions changes the composition of the oils, which may have caused the absence of antimicrobial activity in this case.

Table 2 summarizes the results identified in the work, demonstrating species, micro-organisms against which the antimicrobial activity was reported (or disproved), and the occurrence of positive results in *in vivo* tests and the main action mechanisms in the organism of the animals.

Table 2. Species of plants tested, sensitive and resistant microorganisms, activity as a performance enhancer and primary action of essential oils in the body *Continuation*

Species	Antimicrobial activity [1]	No activity[2]	*in vivo* Activity	Principal action mechanisms
Ginger	*	*Salmonella*	Yes	Performance; carcass yield; Intestinal histomorphometry
Peppermint	*	*	No	Negative results
Lemon	*	*	Yes	Performance; carcass yield
Bay	*	*	Yes	Performance; carcass yield
Basil	*	*	Yes	Performance; carcass yield
Oregano	*Salmonella, E. coli, Clostridium, Eimeria, S. aureus*	*	Yes	Performance; carcass yield; cholesterol levels; intestinal morphometry; hinders oxidation and increases succulence of the meat; histointestinal morphometry; nutrient digestibility
Mexican oregano	*E. coli*	*	Yes	Antimicrobial activity
Red pepper	*Clostridium, Eimeria*	*	Yes	Performance
Red bell pepper	*	*	Yes	Performance; avoid meat oxidation
Pennyroyal	*E. coli*	*Salmonella*	Yes	Performance
Sage	*	*Salmonella*	Yes	Hinders meat oxidation
Thyme	*Salmonella, E. coli, Clostridium,*	*	Yes	Performance; cholesterol levels

[1]Microorganisms sensitive to essential oils, cited in the articles selected.

[2]Micro-organisms resistant to essential oils, cited in the articles selected.

[*]Sensitivity not mentioned in selected articles.

[3]Performance and carcass yields were equal or above those with antibiotics.

The principal microorganisms sensitive to the oils were *Salmonella* spp., *E. coli* and *Clostridium*, besides *Eimeria* spp., *Listeria* spp. and *Staphylococcus aureus* (antimicrobial and anticoccidial effect), also mentioned in the selected articles. As to the benefits of the oil use, as well as their already known bactericidal activity, the most frequently reported actions were: effect on performance and carcass yield (in some studies these parameters improved; in others they were only similar to treatments with the performance enhancers and nevertheless the results were considered positive: the aim of the studies was to verify the possibility of substituting antibiotics, so it is enough that the oils do not worsen the characteristics). A decrease in meat oxidation was also reported, as were a positive effect on the intestinal morphometry and microflora; improvement in flavor characteristics and succulence of the meat; reduced cholesterol and increased nutrient digestibility.

Microencapsulation of Oil

Essential oils are very volatile substances. To work around this problem they can be protected by microencapsulation (Peng et al. 2014; Souza et al. 2008). Among the works mentioned with essential oils as additives five cite the technique as the procedure performed or as a suggestion (Barreto et al. 2008; Betancourt et al. 2012a; Rizzo et al. 2010; Scheuermann et al. 2009; Traesel et al. 2011).

Microencapsulation is essentially a process in which microscopic droplets or particles of the substance (in this case the oil) are enclosed by a layer of encapsulating material, creating tiny capsules (Fernandes et al. 2014). It is used for the purpose of masking unpleasant taste and odor or reduce volatilization and reactivity; increasing the concentration and stability of the substance against harsh environmental conditions (light, oxygen, and pH extremes); it lowers costs; increases facility of storage and protects the substance against the more aggressive processing methods (Comunian et al. 2013; Fernandes et al. 2013; Peng et al. 2014, in addition to promoting the slow release of the encapsulated substance (Alvim and Grosso, 2010; Dima et al. 2014), which can contribute to the oils passing through the stomach intact and be released properly in the intestine.

In the process oils are processed from a liquid form into a dry powder or they remain in liquid form, but are more stable. The most widely used techniques for the microencapsulation of essential oils are spray drying (Fernandes et al. 2014; Souza et al. 2008) and coacervation (Comunian et al. 2013; Peng et al. 2014).

CONCLUSION

There are numerous reports in the literature containing residues of antibiotics and/or resistant bacteria in chicken derived products, though not all present results showing that the resistance was generated by the use of growth promoters and not antibiotics for therapeutic use.

The use of phytotherapics as growth promoters has been shown to be an excellent alternative to common antibiotics, whose use has been questioned in recent years. Most of the related research demonstrates the antimicrobial effect and efficacy of the essential oils in poultry diets with results similar to those obtained with the use of antimicrobials, indicating the feasibility of replacing these drugs.

The literature review, although still small, shows very interesting results for use in this type of research and can be used as a basis for conducting further studies in the area.

References

Agostini, P. S.; Solá-Oriol, D.; Nofrarías, M.; Barroeta, A. C.; Gasa, J.; Manzanilla, E. G. 2012. "Role of in-feed clove supplementation on growth performance, intestinal microbiology, and morphology in broiler chicken." *Livestock Science*, 147:1, 113-118.

Alvim, I. D.; Grosso, C. R. F. 2010. "Microparticles obtained by complex coacervation: influence of the type of reticulation and the drying process on the release of the core material." *Ciência e Tecnologia de Alimentos*, 30:4, 1069-1076.

Amerah, A. M.; Mathis, G.; Hofacre, C. L. 2012. "Effect of xylanase and a blend of essential oils on performance and *Salmonella* colonization of broiler chickens challenged with *Salmonella* Heidelberg." *Poultry Science*, 91:1, 943-947.

Barreto, M. S. R.; Menten, J. F. M.; Racanicci, A. M. C.; Pereira, P. W. Z.; Rizzo, P. V. 2008. "Plant extracts used as growth promoters in broilers." *Revista Brasileira de Ciência Avícola*, 10:2, 109-115.

Betancourt, L. L.; Ariza, C. N.; Díaz, G. G.; Afanador, G. T. 2012a. "Effect of different levels of essential oils of *Lippia origanoides* kunth in broiler chicken." *Revista MVZ Córdoba*, 17:2, 3033-3040.

Betancourt, L. L.; Ariza, C. J.; Afanador, G. T. 2012b. "Effects of supplementation with oregano essential oil on ileal digestibility, intestinal histomorphology, and performance of broiler chickens." *Revista Colombiana de Ciencias Pecuarias*, 25:1, 240-251.

Bona, T. D. M. M.; Pickler, L.; B. Miglino, L. B.; Kuritza, L. N.; Vasconcelos, S. P.; Santin, E. 2012. "Oregano, rosemery, cinnamon essential oil and pepper extract to control *Salmonella*, *Eimeria* and *Clostridium* in broiler chickens." *Pesquisa Veterinária Brasileira*, 32:5, 411-418.

Cho, J. H.; Kim, H. J.; Kim, I. H. 2013. "Effects of phytogenic feed additive on growth performance, digestibility, blood metabolites, intestinal microbiota, meat color and relative organ weight after oral challenge with *Clostridium perfringens* in broilers." *Livestock Science*, 160:1, 82-88.

Comunian, T. A.; Thomazini, M.; Alves, A. J. G.; Matos Junior, F. E.; Balieiro, J. C. C.; Favaro-Trindade, C. S. 2013. "Microencapsulation of ascorbic acid by complex coacervation: Protection and controlled release." *Food Research International*, 52:1, 373-379.

Diarra, M. S.; Silversides, F. G.; Diarrassouba, F.; Pritchard, J.; Masson, L.; Brousseau, R.; Bonnet, C.; Delaquis, P.; Bach, S.; Skura, B. S.; Topp, E. 2007. "Impact of Feed Supplementation with Antimicrobial Agents on Growth Performance of Broiler Chickens, *Clostridium perfringens* and *Enterococcus* Counts, and Antibiotic Resistance Phenotypes and Distribution of Antimicrobial Resistance Determinants in Escherichia coli Isolates." *Applied and Environmental Microbiology*, 73:20, 6566–6576.

Dima, C.; Cotârlet, M.; Alexe, P.; Dima, S. 2014. "Microencapsulation of essential oil of pimento [*Pimenta dioica* (L) Merr.] by chitosan/k-carrageenan complex coacervation method." *Innovative Food Science and Emerging Technologies*, 22:1, 203-211.

Emami, K. N.; Samie, A.; Rahmani, H. R.; Ruiz-Feria, C. A. 2012. "The effect of peppermint essential oil and fructooligosaccharides, as alternatives to virginiamycin, on growth performance, digestibility, gut morphology and immune response of male broilers." *Animal Feed Science and Technology,* 175:1, 57-64.

Erhan, M. K.; Bölükbasi, S. C.; Ürüsan, H. 2012. "Biological activities of pennyroyal (Mentha pulegium L.) in broilers." *Livestock Science,* 146:1, 189-192.

Fernandes, R. V. B.; Borges, S. V.; Botrel, D. A. 2013. "Gum arabic/ starch/ maltodextrin/ inulin as wall materials on the microencapsulation of rosemary essential oil." *Carbohydrate Polymers,* 101:1, 524-532.

Fernandes, R. V. B.; Marques, G. R.; Borges, S.V.; Botrel, D. A. 2014. "Effect of solids content and oil load on the microencapsulation process of rosemary essential oil." *Industrial Crops and Products*, 58:1, 173-181.

Figueiredo, L. S.; Bonfim, F. P. G.; Siqueira, C. S.; Fonseca, M. M.; Silva, A. H.; Martins, E. R. 2009. "Effect of harvesting time on phytomass production and essential oil yield in "alecrim-pimenta" (*Lippia sidoides* Cham.)." *Revista Brasileira de Plantas Medicinais*, 11:2, 154-158.

Furtula, V.; Farrell, E. G.; Diarrassouba, F.; Rempel, H.; Pritchard, J.; Diarra, M.S. 2010. "Veterinary pharmaceuticals and antibiotic resistance of *Escherichia coli* isolates in poultry litter from commercial farms and controlled feeding trials." *Poultry Science*, 89:1, 180-188.

Hong, J. C.; Steiner, T.; Aufy, A.; Lien, T. F. 2012. "Effects of supplemental essential oil on growth performance, lipid metabolites and immunity, intestinal characteristics, microbiota and carcass traits in broilers." *Livestock Science*, 144:1, 253-262.

Karadas, F.; Pirgozliev, V.; Rose, S. P.; Dimitrov, D.; Oduguwa, O.; Bravo, D. 2014. "Dietary essential oils improve the hepatic antioxidative status of broiler chickens." *British Poultry Science*, 55:3, 329-334.

Khattak, F.; Ronchi, A.; Castelli, P.; Sparks, N. 2014. "Effects of natural blend of essential oil on growth performance, blood biochemistry, cecal morphology, and carcass quality of broiler chickens." *Poultry Science*, 93:1, 132-137.

Kirkpinar, F.; Ünlü, B. H.; Özdemir, G. 2010. "Effects of oregano and garlic essential oils on performance, carcase, organ and blood characteristics and intestinal microflora of broilers." *Livestock Science*, 137:1, 219-225.

Kirkpinar, F.; Ünlü, B. H.; Serdaroglu, M.; Turp, G. Y. 2014. "Effects of dietary oregano and garlic essential oils on carcass characteristics, meat composition, colour, pH and sensory quality of broiler meat." British Poultry Science, 55:2, 157-66.

Koaiyama, N. T. G.; Rosa, A. P.; Padilha, M. T. S.; Boemo, L. S.; Scher, A.; Melo, A. M. S.; Fernandes, M. O. 2014 "Growth performance and carcass yield of broiler chickens feed diets with blends of phytogenic additives." *Pesquisa Agropecuária Brasileira*, 49:3, 225-231.

Mathlouthi, N.; Bouzaienne, T.; Oueslati, I.; Recoquillay, F.; Hamdi, M.; Urdaci, M.; Bergaoui, R. 2012. "Use of rosemary, oregano, and a commercial blend of essential oils in broiler chickens: *In vitro* antimicrobial activities and effects on growth performance." *Journal of Animal Science*, 90:1, 813-823.

Melo, M. T. P.; Ribeiro, J. M.; Meira, M. R.; Figueiredo, L. S.; Martins, E. R. 2011. "Essential oil content of pepper-rosmarin as a function of harvest time." *Ciência Rural*, 41:7, 1166-1169.

Rizzo, P. V.; Menten, J. F. M.; Racanicci, A. M. C.; Traldi, A. B.; Silva, C. S.; Pereira, P. W. Z. 2010. "Plant extracts in diets for broilers." *Revista Brasileira de Zootecnia*, 39:4, 801-807.

Osman, K. M.; Elhariri, M. 2013. "Antibiotic resistance of Clostridium perfringens isolates from broiler chickens in Egypt." *Scientific and Technical Review of the Office International des Epizooties*, 32:3, 841-850.

Peng, C.; Zhao, S. Q.; Zhang, J.; Huang, G. Y.; Chen, L. Y.; Zhao, F.Y. Chemical composition, antimicrobial property and microencapsulation of Mustard (Sinapis alba) seed essential oil by complex coacervation. Food Chemistry, v. 165, n. 1, p. 560-568, 2014.

Pinotti, R. N.; Paulillo, L. F. O. 2006. "The structure of the network processing companies of birds in the state of Santa Catarina: contractual governance and resource dependence." *Gestão and Produção*, 13:1, 167-177.

Santos, E. C.; Teixeira, A. S.; Freitas, R. T. F.; Rodrigues, P. B.; Dias, E. S.; Murgas, L. D. S. 2005. "Use of growth promoters additives on performance, carcass yield and total intestinal bacteria counts in broiler." *Ciência Agrotécnica de Lavras*, 29:1, 223-231.

Santurio, D. F.; Costa, M. M.; Maboni, G.; Cavalheiro, C. P.; Sá, M. F.; Pozzo, M. D.; Alves, S. H.; Fries, L. L. M. 2011. "Antimicrobial activity of spice essential oils against *Escherichia coli* strains isolated from poultry and cattle." *Ciência Rural*, 41:6, 1051-1056.

Santurio, J. M.; Santurio, D. F.; Pozzatti, P.; Moraes, C.; Franchin, P. R; Alves, S. H. 2007. "Antimicrobial activity of essential oils from oregano, thyme and cinnamon against *Salmonella* enterica sorovars from avian source." *Ciência Rural*, 37:3, 803-808.

Scheuermann, G. N.; Cunha Junior, A.; Cypriano, L.; Gabbi, A. M. 2009. "Phytogenic additive as an alternative to growth promoters in broiler chickens." *Ciência Rural*, 39:2, p. 522-527.

Shiva, C.; Bernal, S.; Sauvain, M.; Caldas, J.; Kalinowski, J.; Falcón, N.; Rojas, R. 2012. "Oregano essential oil (*origanum vulgare*) and ginger dried extract (*zingiber officinale*) as potential growth promoters in broilers." *Revista de Investigaciones Veterinarias del Perú*, 23:2, 160-170.

Silva, M. A.; Pessotti, B. M. S.; Zanini, S. F.; Colnago, G. L.; Nunes, L. C.; Rodrigues, M. R. A.; Ferreira, L.; Nunes, L. C.; Zanini, M. S.; Martins, I. V. F. 2009. "Intestinal mucosa structure of broiler chickens infected experimentally with Eimeria tenella and treated with essential oil of oregano." *Ciência Rural*, 39:5, 1471-1477.

Silva, M. A.; Pessotti, B. M. S.; Zanini, S. F.; Colnago, G. L.; Nunes, L. C.; Rodrigues, M. R. A.; Ferreira, L. 2011. "Essential oil from Brazillian red pepper as an additive in broiler diet." *Ciência Rural*, 41:4, 676-681.

Souza, R. V.; Silva, V. A. 2008. "Implicações do uso de aditivos na alimentação animal: resíduos e barreiras às exportações." Paper presented at the Animal Production Northeast Congress, 5; Symposium Of Northeastern Ruminant Feeding, 11.; Sergipe Animal Production Symposium,1., Sergipe, Aracaju. November 24–27.

Souza, C. R. F., Schiavetto, I. A., Thomazini, F. C. F., Oliveira, W.P. 2008. "Processing of Rosmarinus officinalis line extract on spray and spouted bed dryers." *Brazilian Journal of Chemical Engineering*. 25:01, 59-69.

Thibodeau, A.; Quessy, S.; Guévremont, E.; Houde, A.; Topp, E.; Diarra, M.S.; Letellier, A. 2008. "Antibiotic resistance in Escherichia coli and Enterococcus spp. isolates from commercial broiler chickens receiving growth-promoting doses of bacitracin or virginiamycin." *The Canadian Journal of Veterinary Research*, 72:1, 129-136.

Traesel, C. K.; Lopes, S. T. A.; Wolkmer, P.; Schmidt, C.; Santurio, J. M.; Alves, S. H. 2011. "Essential oils as substitutes for antibiotic growth promoters in broilers: soroproteins profile and lipid peroxidation." *Ciência Rural*, 41:2, 278-284.

Wareth, A. A. A.; Kehraus, S.; Hippenstiel, F.; Sudekum, K. H. 2012. "Effects of thyme and oregano on growth performance of broilers from 4 to 42 days of age and on microbial counts in crop, small intestine and caecum of 42-day-old broilers." *Animal Feed Science and Technology*, 178:1, 198-202.

Wouafo, M.; Nzoankeu, A.; Kinfack, J. A.; Fonkoua, M. C.; Ejenguele, G.; Njine, T.; Ngandj, A. 2010. "Prevalence and antimicrobial resistance of Salmonella serotypes in chickens from retail markets in Yaounde (Cameroon)." *Veterinary Microbiology*, 16:2, 171-176.

In: Essential Oils
Editor: Miranda Peters

ISBN: 978-1-63484-351-5

Chapter 11

ISHPINK, *OCOTEA QUIXOS* (LAM.) KOSTERM. HISTORY, TRADITIONAL USES, CHEMICAL, PHARMACOLOGICAL PROPERTIES AND THE ECONOMIC POTENTIAL OF ITS ESSENTIALS OILS PRESENT WITHIN THIS AMAZONIAN SPECIES

Paco Noriega, PhD
Centro de Investigación y Valoración de la Biodiversidad,
Universidad Politécnica Salesiana,
Quito, Ecuador

ABSTRACT

The Ishpink tree, otherwise referred to as amazon cinnamon, *Ocotea quixos*, is a plant species that is part of the Lauraceae family. Ishpink is a tree, which is endemic to the Amazonian regions of Ecuador, Colombia and Peru. During the Spanish conquest, rumors of a legendary place full of gold and spices made the conquistadors venture into the impenetrable Amazonian forest. Due to the plant's described properties, it is extremely likely this was the species of legend.

Ishpink is an extremely aromatic plant species. Essential oil can be extracted from the bark, calyx and leaves. Various studies show chemical diversity within the oils, depending on which part it has been extracted from. Through these recent scientific studies, the plant's pharmacological properties are becoming appreciated.

Ecuadorian culinary practices utilize the plant's calyx as an ingredient in the preparation in one of the country's most iconic drinks "colada morada". This is a beverage saved for celebration during the day of the dead.

Recently, Ecuadorian community organizations such as Fundación Chankuap Recursos para el Futuro, have created small centres for the extraction and processing of Ishpink's essential oils. This approach is utilizing the plant's unique properties, translating them into a direct economic benefit for the local community.

The following research analyzes botanical, biological and chemical properties, historical and traditional uses, and potential economic and environmental benefits to the local communities through its sustainable exploitation.

HISTORICAL ASPECTS

"In the provinces that border the eastern diocese of Quito, and are yet to pacify, there is a copious growth of very large trees we call cinnamon. What is most appreciated of these trees are their blossoms; that taste and smell very much like cinnamon, but it is not such a vivid and tasty like cinnamon of East India" (Cobo 1964).

The modern discovery of America by Christopher Columbus occurred because of the belief that travelling east would lead to the territories of India, China and Japan (Philis 1993). In this way, trade with Asia would be developed more easily without the limitations imposed by the Silk Road and land travel, as well as the circumnavigation of Africa. Among the most important products were spices like ginger, cinnamon, jasmine, patchouli and nutmeg (Chaudrhi and Jain 2009).

After Columbus' voyages, the kingdom of Spain began a rapid conquest of territories in Central America and South America. Colonies were established and cities were founded in the most diverse regions of the continent. In the year 1534, the city of San Francisco de Quito was founded and slowly grew, nestled in the middle of the Andes Mountains. It is from here that in 1540 Francisco Pizarro and Francisco de Orellana organized an expedition to find regions rich in spices, the nearly mythical land they called "Tierra de la Canela ("Land of Cinnamon") (Zarate 1947, Fernandez de Oviedo 1959). One of the fortuitous events of this expedition was the navigation and discovery of the Amazon River. In colonial times the Spaniards referred to the Amazon region as an area rich in tobacco for snuff, cocoa, timber, sugar, fragrant resins, medicinal plants and land (Cristobal de Acuna 2009).

One species that certainly contributed to the legend of the existence of the so-called "Tierra de la Canela" was *Ocotea quixos*, also known by the Amazonian natives as Ishpink or Ishpingo (Martinez 2006). Despite the notoriety of this mythical place in the various expeditions and chronicles of the Spanish conquest, the Ecuadorian Amazon region was unexplored for centuries because of its difficult access, warm and humid climate, and ferocity of the various tribes that inhabited it.

The Amazon forests of Ecuador were sparsely settled and this vast territory served as a refuge to several indigenous groups. The greatest changes came since the 1960s, when the exploitation of oil changed the Amazon radically. However, it had the capacity of refocusing our context and putting us back in touch with the myths and legends, namely the originator of all these legends: Amazonian Cinnamon. ous herbs.

Isphingo has traditionally been used for its therapeutic, medicinal and gastronomic properties: it is an appetizer, an eupeptic, and a local anesthetic (Naranjo et al. 1981). Due to its aromatic properties, there is evidence that it has been used since the pre-Columbian era (Naranjo 1969; Star 1990) and is one of the basic condiments in preparing a concoction known as "colada morada", consumed traditionally in Ecuador in celebration of the "Santos difuntos" ("Day of the Dead") (Ulloa 2006).

BOTANY AND ECOLOGY

Ishpink tree, *Ocotea quixos,* is a perennial species belonging to the Lauraceae family, native to South America (Grandtner and Chervete 2013). It tends to grow to the southeast of

Colombia (Camacho, et al. 2002), the south-central Amazon region of Ecuador (Noriega and Dacarro 2008) and the Hamboyacu Altos area in Peru (Peru Progress Report 2012).

The species has been listed as other various genres such as Nectandra, or Licaria Laurus; however it is currently widely accepted in its designation as *Ocotea quixos* (Lam.) Kosterm (Grandtner and Chervete 2013).

Its botanical classification is as follows:

Class: Equisetopsida C. Agardh
Subclass: Novak Magnoliidae ex Takht
Superorder: Takht Magnolianae
Order: Laurales Juss. Ex Bercht. and J. Presl
Species: Ocotea Aubl.
Spice: *Ocotea quixos*. (Tropics 2015)

On average a tree reaches its maturity at six years, with heights ranging between 3 and 6 meters, Figure 1: presents lauroid leaves, ligulate flower stamens and persistent flower challises with six sepals, flowering and fruiting adult trees occur every two years.

Figure 1. *Ocotea quixos* tree.

The leaves, buds and bark have a characteristic smell of cinnamon (Sacchetti et al. 2006), and the essential oil can be obtained from either of these parts. In Colombia, the species are in the endangered category (EN), according to the cataloging of the IUCN (Lopez et al. 2007).

The over-exploitation of resources, and the degradation of habitats have led to the endangerment of the species. In Ecuador, although the species is not listed as endangered (Leon-Yanez, et al., 2011), it grows within the Amazonian environment, which is slowly being destroyed. There is no doubt that the plant could soon enter the endangered category.

CHEMICAL COMPOSITION

Most detailed chemical studies have focused on the essential oils found in the chalices and leaves of this plant species. In the essential oil of the chalices, the main components are: o-methoxycinnamaldehyde, cinnamic acid and methyl cinnamate (Naranjo et al., 1981). Other research highlights the presence of trans-cinnamaldehyde, methyl cinnamate, 1-8 cineole + limonene and p-cymene (Bruni et al. 2004).

The essential oils from the leaves are mostly composed of β-caryophyllene, cinnamyl acetate, sabinene, geranial, and trans-cinnamaldehyde (Sacchetti et al. 2006).

Another research highlights the presence of β-caryophyllene, humulene and eremofilene (Noriega and Dacarro 2008).

So far, the chemical composition of essential oil from the bark of *O. quixos* has not been found in the literature studies.

The following highlights the most important components and their concentrations of essential oils from *O. quixos*, Table 1 and Figure 2.

Table 1. Compounds from essential oils by *Ocotea quixos*

Compound	Calyces		Leaves	
	Naranjo 1981	Bruni 2004	Sacchetti 2006	Noriega 2008
o-methoxycinnamaldehyde	-			
cinnamic acid	-			
methyl cinnamate	-	21.65%		7.21%
trans-Cinnamaldehyde		27.91%	5.1%	3.42%
1-8 cineole + limonene		8.09%		
p-cymene		4.82%		
β-selinene				
cinnamyl acetate			11.4%	4.73%
sabinene			7.6%	
geranial			5.6%	
β-caryophyllene			15.1%	19.02%
humulene				14.32%
eremofilene				11.40%

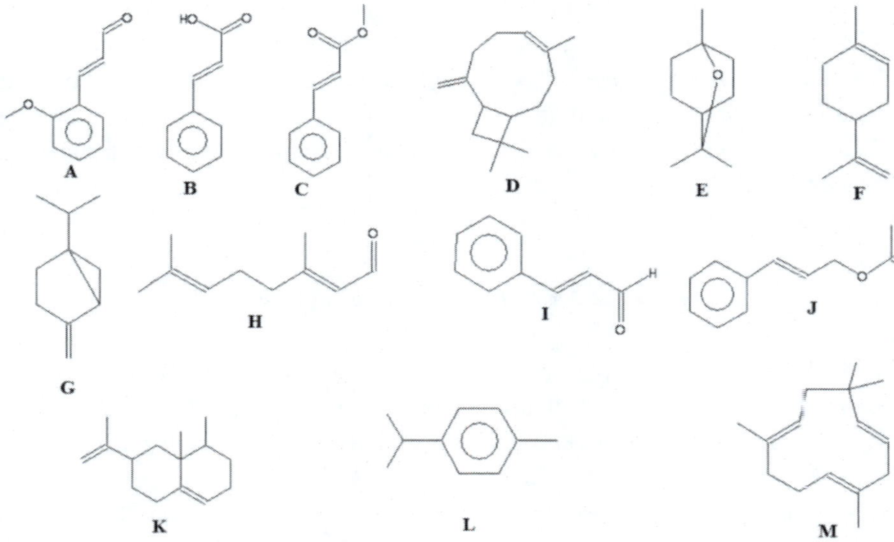

Figure 2. More abundant molecules for essential oils from *O. quixos*. A: *o*-methoxycinnamaldehyde, B: cinnamic acid, C: methyl cinnamate, D: *β*-caryophyllene, E: 1-8 cineole, F: limonene. G: sabinene, H: geranial, I: trans-Cinnamaldehyde, J: cinnamyl acetate, K: eremofilene, L: p-cymene, M: humulene.

Bio-Activity Studies

Antimicrobial and Antifungal Properties

Oils from both chalices and leaves present antimicrobial and antifungal activity. The oil from the calyces presents a very interesting activity, particularly against bacteria such as *S. aureus* (0.12 mg/ml MIC) and *P. aeruginosa* (MIC: 0.049 mg/ml). Other relevant results are noted against yeasts such as *S. cerviseae* (MIC: 0.024 mg/ml) and *C. albicans* (MIC: 0.024 mg/ml) (Bruni, et al., 2004). The oil extracted from the leaves showed increased activity against strains such as *S. pyogenes* (MIC: 0.39.mu.l/ml) and *S. mutans* (MIC: 0.1 .mu.l/ml) (Noriega and Dacarro 2008).

Antioxidant Activity

Studies of antioxidant activity in the oil from the chalice took place through DPPH (free radical scavenging) using the β-carotene test method, which directly evaluates the antioxidant activity. The results were compared to the references used: *T. vulgaris* (natural reference) and Trolox (chemical reference) (Bruni et al. 2004).

Other Studies of Biological Activity

Specific studies for the chalice assess and denote its potential as a potential antiplaletelete and antithrombotic (Ballabeni, et al., 2007), using in vitro methodolcgies: Platelet aggregation, clot retraction, relaxation and aorting ring binding assay; and using in vivo methodologies (preclinical testing), show important activity for acute pulmonary thromboembolism and Bledding test. A recent study reveals important results of both in vitro anti-inflammatory activity by the methods of 774 LPS-stimulated macrophages and forskolin-stimulated SK-N-MC cells; and in vivo methods for rat paw edema and trials in gastric lesions (Ballabeni, et al. 2009).

Ocotea Quixos Potential

All oils of ishpingo have very pleasant aromas with cinnamon notes, however, the chalice oil is particularly aromatic. Because of this, ishpingo is seen as a very promising species, particularly for the perfume and cosmetic industry; equally important is its gastronomic value as an spice and flavoring agent (Muñoz-Acevedo, et al. 2015).

In Ecuador, the social foundation Fundación Chankuap Recursos para el Futuro is the only one dedicated to the extraction and marketing of oils from both chalices and leaves. It also offers several cosmetic products such as soaps and oils prepared for aromatherapy (Chankuap.org 2015). The foundation ensures and promotes a sustainable production implementing good collection practices in the Shuar, Achuar and Colono-Mestizo communities in the provinces of Morona Santiago and Pastaza, and thus the conservation and protection of the species (Chankuap. org 2015).

Essential oils from the ishpink, given their extraordinary aromatic characteristics and multiple chemical and pharmacological studies, make this species extremely important. It has high commercial potential for the pharmaceutical, cosmetic and food industries. The Andean Bio website developed by the Ministry of environment of Ecuador (MAE) and the Corporation for the Promotion of Exports and Investments (CORPEI) mention it as one of the most interesting resources for exploitation (Andean Bio 2015).

Nearly 500 years after the first expedition of Pizarro and Orellana to find the "Tierra de la Canela", the plant that originated this myth and legend continues to be relatively unknown to the outside world. However, its potential promotes continuous research and development of pharmacological studies to understand the behavior of its oils, particularly those extracted from its bark. It is important to promote the protection of the forests in Ecuador and the many flora and fauna species that inhabit them. Given the value of *O. quixos*, its protection takes precedent, as it can be an extremely important resource and income for the many Amazonian communities and indigenous people.

References

Ballabeni, V., Tognolini, M., Bertoni, S., Bruni, R., Guerrini, A., Rueda, G. M., Barocelli, E.: Antiplatelet and antithrombotic activities of essential oil from wild *Ocotea quixos* (Lam.) Kosterm.(Lauraceae) calices from Amazonian Ecuador, 2007, *Pharmacological research*, 55 (1), 23 - 30.

Ballabeni, V., Tognolini, M., Giorgio, C., Bertoni, S., Bruni, R., Barocelli, E.: Ocotea quixos Lam. essential oil: In vitro and in vivo investigation on its anti-inflammatory properties, 2010, *Fitorepia,* 81 (4), 289 - 295.

Biocomercio andino/MAE-CORPEI: Database of flora marketed in Ecuador. 2015/10/22. Available from: URL: http:// www.biocomercioecuador.ec/recursos/base.

Bruni, R., Medici, A., Andreotti, E., Fantin, C., Muzzoli, M., Dehesa, M., Romagnoli, C., Sacchetti, G.: Chemical composition and biological activities of Ishpingo essential oil, a traditional Ecuadorian spice from *Ocotea quixos* (Lam.) Kosterm.(Lauraceae) flower calices, 2004, *Food chemistry,* 85 (3), 415 - 421.

Camacho, R. L., López, D. C, and Hernández, R. R.: Identification manual of wooden spices objetc of marketing in the colombian amazon. 1ª Ed. Bogotá-Colombia: Instituto Amazónico de Investigaciones Científicas: 2002.

Chankuap.org/Fundación Chankuap Recursos para el Futuro: Cosmetics catalog online ishpink. 2015/10/12. Available from: URL: http:// chankuap.org/catalogo/cosmeticos/ linea-ishpink/

Chankuap.org/Fundación Chankuap Recursos para el Futuro: Utilization of ishpink, manual of good practice collection. 2015/10/15. Available from: URL: http://chankuap.org/wp-content/uploads/2014/03/Manual-de-buenas-practicas-de-la-Ishpink.pdf.

Chaudhri, S. K. and Jain, N. K.: History of cosmetics. *Asian Journal of Pharmaceutics,* 2009, 3 (3), 164 - 167.

Cobo, B.: New word history, [1653]. Madrid-España: Biblioteca de Autores Españoles, t. Atlas; 1964.

De Acuña, C.: New discovery of the great amazon river. 1ª Edición. Madrid-España: Editorial Iberoamericana; 2009.

Estrella, E.: America bread. Ethnohistory of the aboriginal food of Ecuador. 3ª edición. Quito-Ecuador: Ed. Abya-Yala; 1990.

Fernadez de Oviedo, G.: General and Natural History of the Indias, Vol I. Madrid-España: Ediciones Atlas; 1959.

Grandtner, M. M, and Chevrette, J.: Dictionary of Trees, Volume 2: South America: Nomenclature, Taxonomy and Ecology. Boston, Mass, USA: Academic Press; 2013.

León-Yánez, S., R. Valencia, N., Pitman, L., Endara, C., Ulloa Ulloa and H. Navarrete.: Red book of the endemic plants from Ecuador. 2ª edición. Quito-Ecuador: Publicaciones del Herbario QCA, Pontificia Universidad Católica del Ecuador: 2011.

López, D., Salinas, N., Colorado, J., Duarte, L., Valencia, S., Hernández, R., Martínez, R., Aguilar, L., Segura, R.: Red book of the Colombian plants. Volumen 4, Wooden species threatened: Primera parte. Bogota-Colombia: Instituto de Estudios Amazónicos. 2007.

Martínez, C. E. C.: Medicinal plants of the ecuadorian andes. Botánica Económica de los Andes Centrales, 2006, 285-293.

Muñoz-Acevedo, A; Torres, E. A.; Gutiérrez, R. G.; Cotes, S. B.; Cervantes-Díaz, M.; Tafurt-García, G.: Some Latin American Plants Promising for the Cosmetic Perfume and Flavor Industries. In: Teixeira-Duarte, M. C.; Rai M. Therapeutic Medicinal Plants from Lab to the Market. Sao Paulo-Brazil: CRC Press; 2015; 279 - 331.

Naranjo, P.; Kijjoa, A.; Giesbrecht, A. M. and Gottlieb, O. R.: *Ocotea quixos*, American cinnamon. *Journal of ethnopharmacology*, 1981, 4(2), 233 - 236.

Noriega, P., and Dacarro, C.: Leaf oil of Ocotea quixos (Lam.) Kosterm.: antimicrobial and antifungal activity, 2008, *La Granja*, 7(1), 3 - 8.

Phillips, W. D., and Phillips, C. R.: The Worlds of Christopher Columbus. Cambridge-UK: Cambridge University Press; 1993.

PUR Projet/Alto Samboyacu: Peru progress report (2012), Alto Shamboyacu project San Martin Region. 2015/09/15. Available from: URL: http:// www.purprojet.com/en/ project/alto-shamboyacu/

Sacchetti, G., Guerrini, A., Noriega, P., Bianchi, A. and Bruni, R. (2006). Essential oil of wild Ocotea quixos (Lam.) Kosterm.(Lauraceae) leaves from Amazonian Ecuador. *Flavour and fragrance journal*, 2006, 21(4), 674 - 676.

Ulloa, C.: Aromas and Andean flavors. Economic Botany of the Central Andes. Universidad Mayor de San Andrés La Paz, 2006, 313, 328.

Tropicos/ Kostermans, A. J. G. H. Ocotea quixos. (Lam) Kosterm. 2015/09/12. Available from: URL: http://www.tropicos.org/Name/17805788.

Zárate, A.: Primitive historians by the Indias, Vol II. Madrid-España: Ediciones Atlas; 1947.

INDEX

A

D

E

F

G

H

I

J

K

L

M

N

O

P

Q

R

S

T

U

V

W

Y